CHEMISTRY RESEARCH AND APPLICATIONS

GRAPHENE

MECHANICAL PROPERTIES, POTENTIAL APPLICATIONS AND ELECTROCHEMICAL PERFORMANCE

CHEMISTRY RESEARCH AND APPLICATIONS

Additional books in this series can be found on Nova's website under the Series tab.

Additional e-books in this series can be found on Nova's website under the e-book tab.

CHEMISTRY RESEARCH AND APPLICATIONS

GRAPHENE

MECHANICAL PROPERTIES, POTENTIAL APPLICATIONS AND ELECTROCHEMICAL PERFORMANCE

BRUCE T. EDWARDS
EDITOR

nova publishers
New York

For permission to use material from this book please contact us:
Telephone 631-231-7269; Fax 631-231-8175
Web Site: http://www.novapublishers.com

Additional color graphics may be available in the e-book version of this book.

Library of Congress Cataloging-in-Publication Data

Graphene : mechanical properties, potential applications and electrochemical performance / [edited by] Bruce T. Edwards.
pages cm
Includes bibliographical references and index.
ISBN 978-1-62948-795-3 (hardcover)
1. Graphene. 2. Graphene--Industrial applications. I. Edwards, Bruce T., editor of compilation.
QD181.C1G74 2014 662'.92--dc23 2013046312

Published by Nova Science Publishers, Inc. † New York

CONTENTS

PREFACE

Graphene has drawn considerable scientific and commercial interest thanks to its unique structure, including being single-atom thick, strictly two-dimensional and highly conjugated, which results in some superb electric, optical, mechanical and thermal properties. In this book the authors present current research in the application of graphene in sensing technology; the adsorption of hexavalent chromium from aqueous solutions by graphene modified with cetyltrimethylammonium bromide; graphene-non-noble metal hybrid nanomaterials as advanced electrocatalysts; the application of graphene in mass spectrometry; the incorporation of graphene into direct-patternable transparent conducting oxide thin films; graphene and related nanomaterials for environmental remediation; and configurations of structural defects in graphene and their effects on its transport properties.

Chapter 1 – Graphene has drawn considerable scientific and commercial interest thanks to its unique structure, including being single-atom thick, strictly two-dimensional and highly conjugated, which results in some superb electric, optical, mechanical and thermal properties. These properties combined with the inherent benefits of the nanocarbon materials make graphene a promising material for the applications in electronics, quantum physics, novel materials and chemistry. In this chapter the authors mainly discuss the recent advance of graphene-based materials in sensing technology including chemical sensors, biosensors, immunosensors for the application in security, enviromental safety and various diseases detection and analysis. The techniques for the preparation of graphene, such as mechanical exfoliation, thermal deposition, chemical vapor deposition (CVD), unzipping CNT, reduction of graphite oxide, particularly the matrix-assisted direct exfoliation of graphene from graphite are briefly introduced in this chapter.

Chapter 2 – In this study, cetyltrimethylammonium bromide was chosen to modify graphene which was prepared using a modified Hummers' method. The characteristics of graphene and modified graphene were characterized by X-ray diffraction, fourier transform infrared spectrum, X-ray photoelectron spectroscopy, transmission electron microscopy and scanning electron microscopy. The effect factors including pH, contact time, temperature and dosage on the adsorption properties of Cr(VI) onto graphene and modified graphene were investigated. Batch experiments were conducted to evaluate the adsorbance of Cr(VI) from aqueous solution using graphene and modified graphene as the adsorbent under different conditions. The results revealed that the optimal pH for the adsorption was about 2, and the best suitable temperature was at 293 K. The adsorption processes were rapid within the first 5 min and reached equilibrium in about 40 min. The adsorption kinetics fitted well with pseudo-second-order model. The adsorption capacity of Cr(VI) on modified graphene inferred from the Langmuir model was 21.57 mg/g at 293 K. The thermodynamic parameters indicated that the adsorption of Cr(VI) onto modified graphene was an exothermic and spontaneous process.

Chapter 3 – Graphene, as the fundamental 2D carbon structure with exceptionally high crystal and electronic quality, has emerged as a rapidly rising star in the field of material science. Its discovery in 2004 led to an explosion of interest in the study of graphene with respect to its unique physical, chemical, and mechanical properties, opening up a new research area for multidisciplinary fields, and aiming for wide-ranging and diversified technological applications. On the other hand, due to the large surface area and high catalytic activity, nanomaterials exhibit promising applications in fuel cells as advanced electrocatalysts. In this chapter, the authors highlight recent advances in the development of graphene-non-noble metal nanomaterials from the standpoint of electrocatalysts. The unique properties of graphene will be discussed firstly. The exciting progress related to graphene-based nanomaterials in electrocatalysts in recent years, including graphene or its derivative (heteroatom doped graphene, etc.), and graphene-supported metal non-noble nanoparticles are summarized in detail.

Chapter 4 – This chapter reviews the up-to-date research about the application of graphene and its relatives in mass spectrometry (MS). Due to its large specific surface area, large delocalized π-electrons, thermal conductivity, stability and rich interaction chemistry, graphene has been widely used in MS study. Graphene and graphene-based materials were applied as very effective matrices for many kinds small molecules in matrix-assisted laser desorption/ionization (MALDI) MS analysis. Many advantages of this novel matrix have been proved, which included: low interference ions from matrix itself, good reproducibility, high salt tolerance, less accompanied reduction and so on. The unique properties of graphene also make it a superior sorbent used in solid-phase extraction (SPE). Further development of on-line SPE methods based on graphene coupled with MALDI-MS, GC-MS and LC-MS greatly simplifies the analytical procedure for complex samples and makes the high-throughput and automatic analysis possible. And its potential in rapid analysis of environmental contaminants has been gradually revealed by more and more research groups. Their applications as a platform in proteolysis for the rapid identification of proteins are also presented here. In addition, laser ablation mass spectra of graphene and graphene oxide have been studied. Graphene was also found to be a unique precursor for the generation of large-sized carbon cluster anions in the gas phase.

Chapter 5 – Transparent metal oxide thin films, such as zinc oxide (ZnO) and tin dioxide (SnO_2) thin films, have been increasingly studied due to their high optical transparency with electrical conducting property, high chemical durability and low cost. An incorporation of graphene into direct-patternable transparent metal oxide thin films was carried out to enhance the electrical conducting property. Metal precursors, solvent, and photosensitive additive were used for the synthesis of stock solutions. After addition of graphene, the solution was spin-coated and the coated film was exposed to ultraviolet (UV) light of 365 nm. Then, the UV exposed film was washed by solvent and the direct-patterned film was annealed to form a desired pure phase system. From the results of UV-vis spectra, X-ray diffractometer, Hall effect measurement and photoelectron spectroscopy of the graphene incorporated metal oxide thin films, the authors can deduce the effect of graphene in transparent metal oxide thin films.

Chapter 6 – It has been widely acknowledged that finding solutions to environmental problems that we are facing now has become a top priority. Environmental problems are highly complex and often vary in scale ranging from local to global dimensions. The advent of graphene revolution in recent years has provided immense scope and opportunities for environmental remediation. Graphene, and its related materials such as graphene oxide, reduced graphene oxide and their nanocomposites can all function as adsorbents and catalysts for the effective removal and degradation of recalcitrant organic pollutants (e.g., dyes, phenolic compounds, polycyclic aromatic hydrocarbons), toxic inorganic contaminants (e.g., heavy metals) and harmful biological substances (e.g., bacteria, microcystins), present invariably in the aquatic environment. Graphene-based materials are also popular in decontaminating industrial waste gaseous streams containing various air pollutants such as NH_3, H_2S, SO_x, NO_x, etc. This chapter aims at bringing together the current knowledge available in the development and use of graphene and its related nanomaterials for environmental clean-up. This synthesis of advanced knowledge in graphene, an emerging field of interest, will help in fuelling further innovations in the development of novel graphene-based nanomaterials for environmental pollution control and abatement, a topic of current global interest.

Chapter 7 – The chapter combines analytical (statistical-thermodynamic and kinetic) with numerical (Kubo–Greenwood-formalism-based) approaches used to ascertain an influence of the configurations of point (impurities, vacancies) and line (grain boundaries, atomic steps) defects on the charge transport in graphene. Possible substitutional and interstitial graphene-based superstructures are predicted and described. The arrangements of dopants over sites or interstices related with interatomic-interaction energies governing the configurations of impurities. Depending on whether the interatomic interactions are short- or long-range, the low-temperature stability diagrams in terms of interaction-energy parameters are obtained. The dominance of intersublattice interactions in competition with intrasublattice ones results in a nonmonotony of orderingprocess kinetics. Spatial correlations of impurities do not affect the electronic conductivity of graphene for the most important experimentally-relevant cases of point defects, neutral adatoms and screened charged impurities, while atomic ordering can give rise in the conductivity up to tens times for weak and strong short-range potentials. There is no ordering effect manifestation for long-range potentials. The anisotropy of the conductivity along and across the line defects is revealed and gives rise in the conductivity of graphene with correlated line defects as compared with the case of random ones. Simultaneously correlated (and/or ordered) point and line defects in graphene can give rise in the conductivity up to hundreds times vs. their random distribution. On an example of different B or N doping configurations in graphene, results from the Kubo–Greenwood approach are compared with those obtained from DFT method.

In: Graphene
Editor: Bruce T. Edwards
ISBN: 978-1-62948-795-3

Chapter 1

APPLICATION OF GRAPHENE IN SENSING TECHNOLOGY

***Jingquan Liu**[*1]**, Aihua Li**[2]**, Na Kong**[1]* ***and Shengyu Feng**[†2]*

[1]College of Chemistry, Chemical and Environmental Engineering, Laboratory of Fiber Materials and Modern Textile, The Growing Base for State Key Laboratory, Qingdao University, Qingdao, P. R.China
[2]Key Laboratory of Special Functional Aggregated Materials, Ministry of Education and School of Chemistry and Chemical Engineering, Shandong University, Jinan, P. R. China

ABSTRACT

Graphene has drawn considerable scientific and commercial interest thanks to its unique structure, including being single-atom thick, strictly two-dimensional and highly conjugated, which results in some superb electric, optical, mechanical and thermal properties. These properties combined with the inherent benefits of the nanocarbon materials make graphene a promising material for the applications in electronics,

* College of Chemistry, Chemical and Environmental Engineering, Laboratory of Fiber Materials and Modern Textile, The Growing Base for State Key Laboratory, Qingdao University, Qingdao 266071, P. R.China, Email: jliu@qdu.edu.cn.
† Key Laboratory of Special Functional Aggregated Materials, Ministry of Education and School of Chemistry and Chemical Engineering, Shandong University, Jinan 250100, P. R. China, Email: fsy@sdu.edu.cn.

quantum physics, novel materials and chemistry. In this chapter we mainly discuss the recent advance of graphene-based materials in sensing technology including chemical sensors, biosensors, immunosensors for the application in security, enviromental safety and various diseases detection and analysis. The techniques for the preparation of graphene, such as mechanical exfoliation, thermal deposition, chemical vapor deposition (CVD), unzipping CNT, reduction of graphite oxide, particularly the matrix-assisted direct exfoliation of graphene from graphite are briefly introduced in this chapter.

1. INTRODUCTION

In recent years, nanomaterials have evidenced rapidly advance since the emergence of a two-dimensional material, graphene, which consists of hexagonal configuration with sp^2 bonded carbon atoms. The experimental isolation of individual graphene sheets was first achieved with a simple repeated peeling process by Geim and Novoselov in 2004.[1] Since then, large numbers of scientists have created ingenious methods to synthesize monolayered graphene flakes, including mechanical and matrix assisted exfoliations,[2-4] thermal deposition,[5,6] chemical vapor deposition (CVD),[7,8] unzipping CNT,[9,10] and reduction of graphite oxide.[11,12] Each of these preparation methods has some advantages for specific applications, but also has some drawbacks. For example, graphene prepared by mechanical exfoliation and cleavage exhibits high carrier mobility and low density of defects, therefore it can be used for fabrication of electronics. However, mechanical exfoliation is a time consuming process which can not be applied for large-scale production of graphene. The oxidation-reduction method for producing graphene from graphite can be used for mass production, whereas the as-prepared graphene nanosheets usually retain smaller size, high defects and low electrical conductivity.

Though graphene is the most recently discovered member of the nanocarbon family, it has already been intensively applied in various fields due to its unique chemical and physical properties (such as superior electrical conductivity, excellent mechanical flexibility, large surface area and high thermal and chemical stability). For instance, graphene was explored for the wide application in energy field, attributed to its high conductivity, transparency and ultrathin sheets.[13-16] Owing to the large specific surface area (2630 m^2 g^{-1}),[17] excellent mechanical strength and aromatic–rich structure, graphene can be applied as pollutant adsorbents by strong interaction between graphene surface and small molecules, as well as catalyst or catalytic support for fuels and photo degradation of organics.[18-22] Moreover, graphene plays a crucial role in sensing application which utilize graphene's exceptional electrical properties (e.g., extremely high carrier mobility and capacity), electrochemical properties (e.g., high electron transfer rate), optical properties (e.g., excellent ability to quench fluorescence), structural properties. In this chapter, we place the emphases on graphene-based materials for their applications for the detection of gas, heavy metal ions, protein, glucose and bacteria etc.

2. Preparation of Graphene

Since Geim and Novoselov et al.[1] were credited for the successful preparation of graphene by the now famous Scotch tape method in 2004, many methods have been reported on the synthesis of graphene. Generally speaking, these methods can be classified as exfoliation, thermal decomposition, chemical vapor deposition (CVD), unzipping CNT, thermal reduction and oxidation-reduction etc, which will be described below.

2.1. Mechanical and Matrix-Assisted Exfoliation

Exfoliation, a repeated peeling process to prepare graphene from highly-oriented pyrolytic graphite (HOPG).[1] Using this technique, graphene samples obtained were virtually free of crystal defects, resulting in high carrier mobility. Likewise, Shukla et al.[2] prepared mm-sized single- to few-layer graphene by bonding bulk graphite to borosilicate glass under a particular temperature and voltage, followed by exfoliation to leave single- or few-layer of graphene on the substrate. However, mechanical exfoliation is a time consuming process and cannot be scaled for industrial large-scale production. Therefore, other exfoliation methods have been developed.

It was found that graphene can be exfoliated directly from graphite in organic solvents such as N-methyl-pyrrolidone.[3] The monolayer graphene yield reached 7-12 wt % by further processing. Similar approach was adopted to obtain garphene sheets by sonication of natural graphite crystals in dimethylformamide (DMF).[23] Solution based exfoliation of graphite showed a new window for large-scale production of graphene. Graphene was prepared using a similar process of liquid based exfoliation in water-surfactant solutions with the aid of ultrasound.[4] Another group also reported the exfoliation-reintercalation-expansion of graphite to produce high-quality single-layer graphene sheets (GSs) stably suspended in DMF (Figure 1).[24] Gram-scale carbon nanosheets can be prepared by low-temperature flash pyrolysis of a solvothermal product of sodium and ethanol, followed by mild sonication of the nanoporous carbon product.[25]

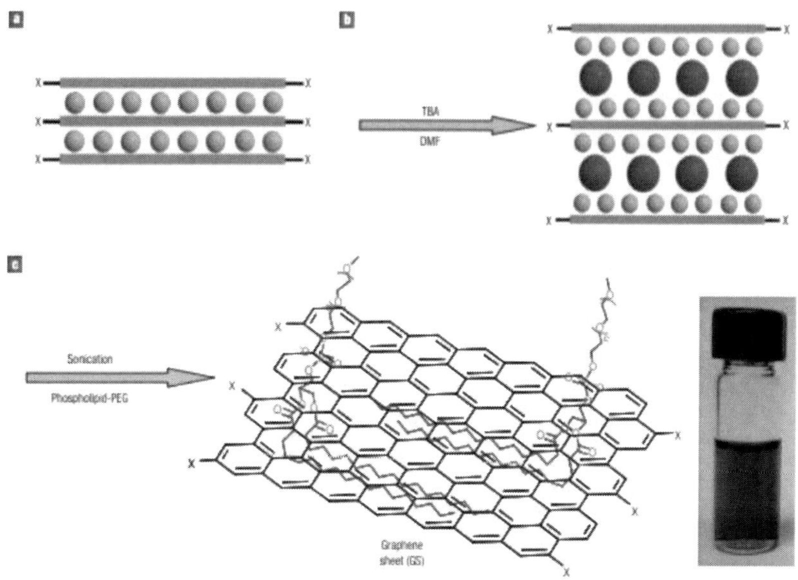

Figure 1. Chemically derived single-layer GS from the solution phase. a), Schematic of the exfoliated graphite reintercalated with sulphuric acid molecules (teal spheres) between the layers. b), Schematic of tetrabutylammonium hydroxide (TBA) (blue spheres) insertion into the intercalated graphite. c), Schematic of GS coated with 1,2-distearoyl-sn-glycero-3-phosphoethanolamine-N-[methoxy(polyethyleneglycol)-5000] (DSPE-mPEG) molecules and a photograph of a DSPE-mPEG / DMF solution of GS.

Moreover, enormous research focused on the preparation of graphene by non-destructive π–π stacking interactions between the aromatic molecules and graphite microplatelets. Jang et al.[26] reported a supercritical fluid (SCF) method to obtain high quality graphene sheets by introducing 1-pyrene sulfonic acid sodium salt (1-PSA) during the SCF exfoliation. This process could result in homogeneously modified graphene nanosheets (GNS) via the π–π interaction between the graphene layer and aromatic ring of the modifier molecules without any damages to the graphene surface and its electronic conductivity. Likewise, single stranded DNAs modified 1-pyrenebutyric acid was also utilized to prepare a water-dispersible single-layered graphene from exfoliation of graphite flakes.[27] Later, the amphiphilic copolymers, consisting of a mono- or multi-pyrene units, were designed to selectively exfoliate graphite powder into single- and double-layer graphene sheets in aqueous or organic media.[28,29] Stable suspensions of graphene nanosheets were also successfully produced by using chemical exfoliation of graphite oxide and then reducing it to graphene.[30]

2.2. Thermal Decomposition

The thermal decomposition of silicon carbide (SiC) is a technique employed for the fabrication and processing of graphene materials. When SiC is heated in ultra-high vacuum (UHV) to temperatures between 1000 °C and 1500 °C, Si will sublimate and leave behind a carbon rich surface. Bergen et al.[5] prepared ultrathin epitaxial graphite films on the Si-terminated (0001) face of single-crystal 6H-SiC by thermal decomposition of SiC. The Si face was treated by oxidation or H_2 etching, followed by being heated via electron bombardment in ultrahigh vacuum to 1000 °C in order to remove the oxide. After deoxygenation, samples were heated to temperature ranging from 1250 °C to 1450 °C for 1-20 min and thin graphite layers were formed. The graphene layers were from 1 to 3 and the thickness of graphene mainly relies on the temperature. Likewise, graphite thin films were synthesized by thermal decomposition and demonstrated the successful growth of single crystalline films down to approximately one graphene layer.[31] Lin et al.[6] fabricated graphene field-effect transistors (FETs) on a 2-inch graphene wafer, made by epitaxially grown on the Si face of high-purity SiC wafer via thermal annealing at 1450 °C and further demonstrated the high potential application of graphene for electronic devices. However, vacuum decomposition of SiC yields graphene layers with small grains. To solve this problem, *ex* situ graphitization of SiC under 900 mbar of argon was performed to prepare monolayer graphene films with much larger domain sizes (more than 50μm in length) and high electronic mobility of 2,000 $cm^2\ V^{-1}\ s^{-1}$.[32] In addition, Juang et al.[33] presented a modified method of thermal decomposition to fabricate millimeter-scaled graphene films on silicon carbide substrates at low temperatures (750 °C).

In a similar approach by Sutter et al.,[34] epitaxial graphene was grown on the (0001) surface of ruthenium (Ru). They found that interaction between the first graphene layer and metal substrate was strong, while the second layer was almost completely detached, showing weak electronic coupling to the metal. This result showed that graphene sheets obtained through this approach retained the inherent electronic structure of graphene. This finding provided a feasible route towards rational graphene synthesis on transition-metal templates. Another independent study[35] has also reported on epitaxial graphene monolayer grown on Ru (0001) covering uniformly the substrate over lateral distances larger than several microns, with periodical ripple and spatial extensibility.

2.3. Chemical Vapor Deposition (CVD)

In addition to the thermal decomposition of SiC, graphene with high structural quality and large area can also be obtained through low-pressure chemical vapor deposition, in which carbon is supplied in gas form and a metal is used as both catalyst and substrate to grow the graphene layer. Somani et al.[7] reported for the first time the synthesis of plane few-layer graphenes (PFLG) from camphor pyrolysis on nickel substrates by a simple, cost-effective thermal CVD method. Graphene sheets prepared by Acro et al.[36] via the CVD method exhibited high transmittance of 80 % in the visible wavelength range. Another recent study[37] has shown that the synthesis of large area, few-layer graphene films on polycrystalline Ni films via ambient pressure CVD, which can be transferred to arbitrary substrates, keeping its electrical properties unchanged. The as-prepared graphene films with 3 nm average thickness transferred onto a glass substrate exhibited high optical transmittance of ~ 90 % in the 500-1000 nm wavelength regime. Similarly, single- or multi-layered graphene sheets with high transmittance, electrical properties and large size have also been successfully prepared by a few other groups.[38,39] The CVD methods was also conducted on the copper substrate to prepare graphene sheets with similar properties.[40,41] It indicated that the process was the surface mediated and self-limiting. A roll-to-roll process has also been developed to produce graphene monolayer up to 30-inch by CVD on copper.[8]

2.4. Unzipping CNT

An elegant method for the fabrication of graphene ribbons and sheets via unzipping CNT with intercalation of lithium and ammonia followed by exfoliation was reported.[9] Kosynkin and coworkers[42] prepared a nearly 100 % yield of graphene nanoribbon (GNR) structures by lengthwise cutting and unraveling of carbon nanotubes (CNTs). In addition, Jiao et al.[10] developed an approach to produce narrow GNRs by controlled unzipping of CNTs by plasma etching of nanotubes partly embedded in a polymer film. It was demonstrated that the preparation of single-layer or multilayer GNRs and GNRs with inner CNT cores depended on the diameter and number of layers of the starting MWCNT and the etching time. The GNRs synthesized by an argon (Ar) plasma etching method had smooth edges and a narrow width distribution (10-20 nm).

2.5. Oxidation-Reduction of Graphite

Oxidation-reduction is the most popular method for the preparation of graphene in large scale. It usually consists of two steps: The first one involves the oxidation of graphite using strong acids and oxidants to afford graphene oxide (GO)[43,44] and the second one will convert GO into graphene in the presence of versatile reductants.[11,45,46] Graphene sheets prepared by the reduction of GO usually contain a few of oxygen-containing functional groups, resulting in further modification with other materials for producing large-scale flexible conductors and conductive devices. Compared with other methods, the chemical reduction approach is a promising one since it can be scalable in production and versatile in realizing abundant chemical functionalization. In addition, it has been demonstrated that the electrical conductivity of GO can be restored close to the level of graphite by chemical reduction.[47,48] Stankovich et al.[11] developed a method to reduce the exfoliated GO sheets in water with hydrazine hydrate ($N_2H_4{\cdot}H_2O$) to afford graphene. In another work, Shin and coworker[12] demonstrated sodium borohydride ($NaBH_4$) could be a more effective reducing reagent than hydrazine (N_2H_4) for reduction of GO. The sheet resistance of GO reduced by $NaBH_4$ was much lower than that of N_2H_4-reduced films. This feature was attributed to the continuous accumulation of nitrogen atoms forming C-N groups in the N_2H_4-reduced films, which behaved as donors compensating p-type hole carriers in reduced graphite oxide, resulting in high sheet resistance.

However, in the reduction of GO using $NaBH_4$, it was found that the interlayer distance was first slightly expanded by forming the intermediate boron oxide complexes, followed by the contraction with the gradual removal of carbonyl and hydroxyl groups. The sheet resistance of conductive graphene films fabricated with a $NaBH_4$-reduced GO was comparable to that of dispersed graphene. Nevertheless, the $NaBH_4$ reducing reagent has a low effect on the removal of epoxy groups and carboxylic acids. Therefore, a dehydration with concentrated sulfuric acid after deoxygenation with $NaBH_4$ was proposed, which was a really simple and effective route to further improve the reduction effect on GO.[45]

Recently, Pei et al.[46] reported a simple but highly-effective reducing method using a strong reagent like hydroiodic acid (HI) to reduce GO films based on the nucleophilic substitution reaction to prepare graphene films without destroying their integrity and flexibility at low temperature. In addition, another recent report on the reduction of GO employing $_{L}$-ascorbic acid (Vitamin C) as the reductant and $_{L}$-tryptophan as the stabilizer was made by Gao et al.[49] The use of biocompounds as reducing reagents for nontoxic and scalable production of graphene was the first time. Other reductants have also been explored to efficiently reduce GO, such as polymers, sulfur-containing compounds and hot strong alkaline solutions (KOH, NaOH).[50-52]

2.6. Other Methods (Thermal, Microwave and UV Induced etc.)

In addition to chemical reduction of GO, some other reduction methods have been explored for preparing graphene sheets. For example, Zangmeister et al.[53] prepared RGO by the reduction of GO with thermal processing at 220 °C, along with the loss of oxygen and formation of new sp^2 C-C bonds. Moreover, a simple yet versatile method for the exfoliation and reduction of GO was also reported, where the reduced graphene oxide materials could be readily prepared within 1 min in microwave oven.[54] The specific capacitance of microwave assisted exfoliated and reduced GO reaches up to 191 F g^{-1} with KOH electrolyte. The well-separated graphene-semiconductor composite sheets were also prepared through UV-induced photocatalytic reduction of GO.[55]

3. Graphene Based Chemical Sensors

Despite its short history, graphene, has attracted a great deal of interest in the development of new sensors due to the high conductivity, good biocompatibility, electronic property and potential applications in chemical sensors. The high-performance of graphene based sensors for the detection of gas, heavy ions and organic molecules were in the discussion below. Furthermore, different kinds of graphene precursors, such as chemically reduced graphene oxide, graphene grown by CVD and graphene nanoribbons and their effects on the device fabrication and sensing performance were also discussed.

3.1. Gases Detection

While not as matured, the graphene-based sensors for several gaseous and vapor analytes were growing, and their electrical response to various chemical species was reported to be qualitatively similar to CNTs.[56] Graphene-based sensor for gas detection, for the first time, was demonstrated by Novoselov and co-workers.[57] Graphene sensor based on micromechanical cleavage of graphite at the surface of oxidized Si wafers detected NO_2 gas by measuring the change of source–drain resistance. The removal of electron density decreased the resistance of the graphene device, while the donation of electron density increased the device resistance. Due to the extremely low intrinsic noise of nearly-defect-free graphene, the high sensitivity was achieved. Therefore, the ultimate detection of a single NO_2 gas molecule was attributed to the high carrier mobility of graphene and the extremely low noise of the device. Several other studies have been reported for the detection of NO_2 with graphene-based devices. For example, Jeong et al.[58] fabricated a flexible room-temperature NO_2 gas sensor by the growth of a vertically aligned CNT array (with CNTs 20 μm in length) on the reduced graphene film surface supported by a polyimide substrate. In comparison with the reduced graphene film, the CNTs/reduced graphene hybrid film showed remarkably enhanced sensitivities with weak N–P transitions, mainly attributed to the high sensitivity of the CNT arrays. Reduced graphene oxide (rGO)-conjugated Cu_2O nanowire mesocrystals were prepared as a NO_2 sensor, which was synthesized by nonclassical crystallization in the presence of GO and o-anisidine under hydrothermal conditions.[59] This rGO conjugated Cu_2O mesocrystal composite combined the advantages of enhanced electronic conductivities of rGO and the interdendritic space within the mesocrystals formed 3D conducting networks. Because the porous three-dimensional (3D) framework structures improved conductivity, the rGO-Cu_2O mesocrystals exhibited a higher sensitivity for the detection of NO_2 gas at room temperature than that of standalone systems of Cu_2O nanowires networks and rGO sheets. Cu_2O nanocrystals modified graphene nanocomposite based sensor was also explored for room temperature H_2S gas sensing with ultrahigh sensitivity.[60] These studies undoubtedly offered the mesocrystal-based nanodevices possibility of detecting various gases. A graphene gas sensor using ozone treatment was also exhibited remarkably enhanced sensing performance such as percentage response, detection limit and response time for the detection of NO_2 gas.[61] It was found that the percentage response of the ozone treated graphene sensor was twofold higher, compared with that of a pristine graphene sensor, when 200 ppm concentration of NO_2 was exposed at room temperature. The response time was significantly improved by a factor of 8.

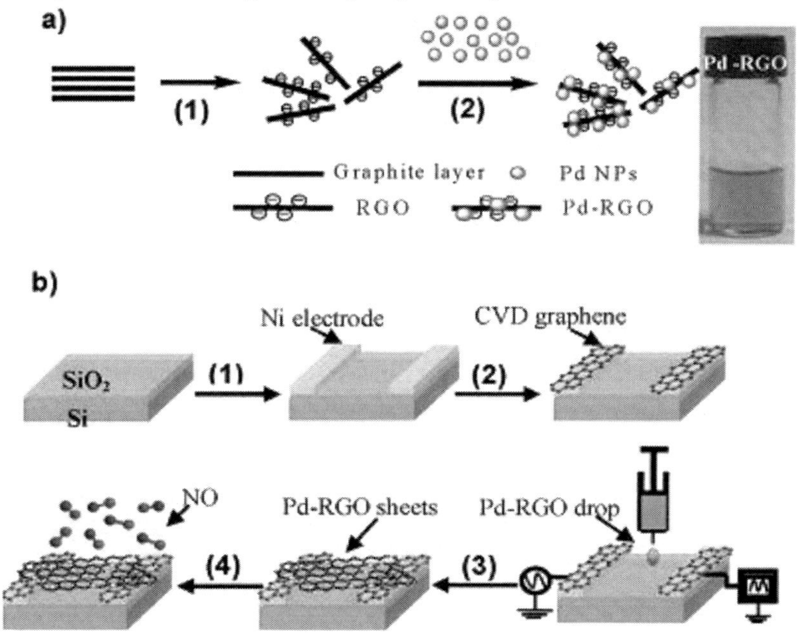

Figure 2. a) Schematic illustration of the process for preparation of Pd-rGO composites: (1) rGO synthesis and (2) Pd decoration of rGO. The inset photograph is the diluted Pd-rGO nanosheet suspension used for ac-DEP. b) Schematics of graphene-Pd-rGO device fabrication and gas sensing test: (1) Ni electrode fabrication, (2) chemical vapor deposition growth of graphene, (3) ac-DEP of Pd-rGO nanosheets, and (4) sensor measurement.

Besides, a flexible and lightweight chemiresistor was developed by Dua and coworkers[62] for the detection of NO_2 with a ultralow LOD (~ 400 ppt). The NO_2 sensor was prepared by inkjet-printing reduced graphene oxide film on a poly-(ethylene terephthalate) (PET) plastic film substrate. Inkjet printing of rGO platelets was achieved for the first time using aqueous surfactant-supported dispersions of rGO powder prepared by ascorbic acid reduction of dispersed graphene oxide. Fowler and coworkers[63] also reported the development of useful chemical sensors using rGO sheets spin-coated on an interdigitated electrode array and preliminary results were presented on the detection of NO_2, NH_3, and 2,4–dinitrotoluene. It was found that the sensor response supported the charge transfer mechanism between the analyte and graphene with a limited role of the electrical contacts. Similarly, chemiresistor-type NO_2 sensors with high sensitivity, selectivity and reversibility were also fabricated based on chemically modified graphene (CMG) including sulfonated rGO (S-G) or ethylenediamine-modified rGO (EDA-G).[64] The responses of CMG based sensors were stronger compared with those of conventional rGO-based sensors, and also showed good linear relationships in wide NO_2 concentration ranges with low detection limits. The work indicated that chemical modification of graphene could improve the materials' performance for gas sensing.

In addition to the NO_2 sensing, graphene based materials were reported for the detection of NH_3 gas. Lu et al.[65] fabricated a reduced graphene oxide sensor for fast, repeatable, room temperature NH_3 gas detection. The authors demonstrated that the rGO field-effect transistors (FET) operated in n-type mode by applying sufficiently positive gate voltage (V_g) exhibited faster instantaneous response and faster recovery for NH_3 sensing in comparison with their performance in p-type mode, which was attributed to the ambipolar transport of rGO and the V_g-induced effects. This V_g -dependent NH_3 sensing facilitated rGO based sensors application for the room temperature gas detection. Another NH_3 sensing device was fabricated based on patterned, vertically oriented graphene sheets obtained from plasma-enhanced chemical vapor deposition (PECVD).[66] The authors suggested the electric field distribution above the substrate material had an important influence on the patterned synthesis of vertical graphene nanosheets. Huang et al.[67] also studied the adsorption of gas molecules (CO, NO, NO_2, O_2, N_2, CO_2, and NH_3) on graphene nanoribbons (GNRs) and found that GNR with armchair-shaped edges could exhibit n-type semiconducting behavior after NH_3 adsorption, while other gas molecules had little effect on modifying the conductance of

GNRs. The result further indicated that NH_3 molecules could be detected out of these gas molecules by GNR based sensor.

Besides the commonly demonstrated NO_2 and NH_3 sensors, graphene based gas sensors were also employed in detection of H_2,[68-70] CO_2,[71] NO[72] and H_2O.[73] For instance, Pt/graphene-like nanosheet based H_2 sensor was reported by Shafiei et al.[68] which was prepared via the reduction of spray-coated graphite oxide deposited on SiC substrates by hydrazine vapor, followed by a circular pad of Pt deposited on top of the rGO film forming the electrical contacts. The authors studied current voltage and dynamic responses of the sensors towards different concentrations of hydrogen gas in a synthetic air mixture at 100 °C. The dynamic response exhibited a voltage shift of 100 mV for 1 % hydrogen in synthetic air at 100 °C. Johnson et al.[69] used palladium (Pd)-functionalized multi-layer graphene nanoribbon (MLGN) networks for the detection of H_2 gas. It was demonstrated that the porous structure and high specific surface area of the nanoribbon networks facilitated efficient functionalization and high sensitivity to H_2 gas at room temperature ($\Delta R/R \sim 55$ % for 40 ppm H_2). A high-performance graphene CO_2 gas sensor was also fabricated by mechanical cleavage.[71] Results showed that the conductance of the graphene gas sensor increased linearly with the concentration of CO_2 gas increasing from 10 to 100 ppm. A graphene-based NO sensor devices fabricated by alternating current dielectrophoresis (ac-DEP) of bulk Pd-rGO nanosheets with CVD-grown graphene electrodes is shown in Figure 2.[72] The novel sensing device exhibited high sensitivity, recoverability and fast responsiveness for the detection of NO gas ranging from 2 to 420 ppb at room temperature.

Recent studies on rGO-based gas sensors were also reported for the detection of several chemical vapors, such as ethanol,[74] methanol [75] and dinitrotoluene (DNT).[63] Jiang et al.[74] firstly produced Al_2O_3/Graphene nanocomposite with GO solution by a one-step, green, facile, low-cost supercritical CO_2 method. The synthesized Al_2O_3/graphene nanocomposite exhibited high catalytic chemiluminescence sensitivity and high selectivity for ethanol. Most recently, an array of rGO-based integrated sensors was fabricated to recognize different alcohols, such as methanol, ethanol and isopropanol.[75] Because of the irregular structure of rGO films at different levels of organization, each rGO-based device in such an array had a unique sensor response, which solved the problem of poor selectivity of conventional rGO-based gas sensors. The high selectivity to discriminate between different alcohols (methanol, ethanol and isopropanol) at a 100 % success rate was demonstrated in their experiments. This study provided great prospect for different gas molecules at the same time.

3.2. Heavy Metal Ions Detection

Since heavy metal ions included in water or food resources have a vital effect on human health, the highly sensitive, rapid, and selective analysis for heavy metal ion detection has been extensively explored by means of electrochemical, optical, and colorimetric methods.[76] Graphene-based electrochemical sensors have been employed to detect and monitor the presence of heavy metal ions such as Pb^{2+}, Cd^{2+} and Hg^{2+}.[77-84] Li et al.[77,78] reported the electrochemical sensor based on the Nafion–graphene nanocomposite film for the determination of Pb^{2+} and Cd^{2+} by anodic stripping voltammetry (ASV). The Nafion-graphene composite film not only exhibited improved sensitivity for the metal ion detections, but also alleviated the interferences due to the synergistic effect of graphene nanosheets and Nafion. Results showed that the stripping current signal was significantly enhanced on graphene electrodes. The linear calibration curves ranged from 0.5 μg L^{-1} to 50 μg L^{-1} for Pb^{2+} and 1.5 μg L^{-1} to 30 μg L^{-1} for Cd^{2+}, respectively. The detection limits (S/N = 3) were around 0.02 μg L^{-1} for Pb^{2+} and Cd^{2+}.

The Nafion–graphene nanocomposite modified electrochemical sensors have also been verified in the water sample determination. Lien et al.[79,80] fabricated the Nafion-modified graphene/carbon nanotube composite deposited with Bi on the screen-printed electrode (Nafion-G/CNT-BiSPE) for the determination of trace Pb^{2+}. To gain a basic understanding for the complicated influences on the current response of lead-ion sensing, five factors, including Bi deposition time, concentration of Nafion, pH, pre-concentration time and pre-concentration potential were investigated. It was found that the pH value and the concentration of Nafion had a significant influence on the sensitivity of Pb^{2+}.[80] When 0.3 % Nafion and the stripping buffer solution with pH = 4.75 were employed, the stripping current response of Pb^{2+} sensing on Nafion-G/CNT-BiSPE could reach the highest value. Furthermore, due to smooth electron pathways and effective exposure of Bi nanoparticles on three-dimensional G-CNT and the negative charge of Nafion, the current response of Pb^{2+} on Nafion-G/CNT-BiSPE was estimated to be over 50 times in comparison with that of a Bi-deposited SPE electrode.[79] The in situ bismuth-modified graphene-carbon paste electrode (Bi-GCPE) was also developed for determining Cd^{2+} and Pb^{2+} levels in an automated flow system.[85] This environment friendly electrode exhibited excellent electro oxidation of Cd^{2+} and Pb^{2+} with a significantly higher peak current for both metal ions compared with the unmodified carbon paste electrode (CPE). Under optimum conditions, the detection limits from Bi-GCPE were 0.07 and 0.04 μg L^{-1} for Cd^{2+} and Pb^{2+}, respectively, and the linear oxidation peak current response for Cd^{2+} and Pb^{2+} was in the range of 0.10 – 50.0 μg L^{-1}. The Bi-GCPE was also applied for the determination of Cd^{2+} and Pb^{2+} in low- (tap water) and high- (sea bass fish and undulated surf clam tissues) matrix complexity samples and the results were satisfying.

Besides, a gold electrode has also been successfully modified with graphene via non-covalent interaction for the analysis of Pb^{2+} and Cu^{2+}.[82] The modified gold electrode was prepared by the first modification with carboxylic acid functionalities, followed by covalent attachment of pyrene, through which graphene nanosheets were then immobilized via non-covalent π–π stacking interaction. (Figure 3) The graphene modified gold electrode exhibited enhanced analytical sensitivity toward Cu^{2+} and Pb^{2+} and good reusability and repeatability were also observed. The graphene modified carbon nanosheets (GMCN) were also explored for sensitive detection of trace amount of Pb^{2+} from an aqueous solution.[83] GMCN was possessed of the advantages of both porous carbon and graphene sheet, which gave rise to the accumulation of Pb^{2+} and electronic transport properties, and GMCN could make a good balance between electrical conductivity and surface area, which was crucial for an excellent sensitivity and detection limit. Except for the above advantages, graphene modified porous carbon with good chemical stability could be applied in harsh condition, for example, strong acid or base.

In reported graphene based sensors for the detection of Pb^{2+}, gold nanoparticles (AgNPs) were also employed due to its excellent electrical conductivity and admirable capability of quenching fluorophores such as organic fluorescent tags.[81,86] Lu et al.[86] described a simple and fast strategy for the fabrication of a glassy carbon electrode for the first time by single-step electrodeposition of graphene (GR), gold nanoparticles (AgNPs), and chitosan (CS) directly from a solution containing graphene oxide, tetrachloroauric acid, and chitosan. The prepared GR–AuNPs–CS was firstly combined with bismuth–film electrode for the trace analysis of Pb^{2+} by differential pulse anodic stripping voltammetry and successfully applied to the detection of Pb^{2+} in spiked samples of river water. Results showed that a linear relationship existed between electrical current and the concentration of lead (II) ion in the range between 0.5 to 100 μg L^{-1}, with a detection limit of 1 ng L^{-1} (at an SNR of 3) under the optimum conditions. A novel platform for effective "turn-on" fluorescence sensing of Pb^{2+} in aqueous solution was also explored based on AuNP-functionalized graphene taking advantage of the quenching capability of AuNPs as well as the optical property of graphene.[81] In this novel "turn-on" fluorescence sensor, graphene was used as fluorescence creating source, and the interaction between graphene and AuNPs could cause the fluorescence intensity to decrease/quench. Interestingly, upon the addition of Pb^{2+}, the fluorescence of graphene could reappear and increase, which was attributed to the fact that Pb^{2+} could accelerate the leaching rate of the AuNPs on graphene surfaces in the presence of both thiosulfate ($S_2O_3^{2-}$) and 2-mercaptoethanol (2-ME) (Figure 4). Therefore this offered an ideal alternative for selective detection of Pb^{2+}. It was found that the relative fluorescence intensity showed good linearity versus logarithm concentration of Pb^{2+} in the range of 50 – 1000 nM (R = 0.9982), and a detection limit of 10 nM with the optimum conditions.

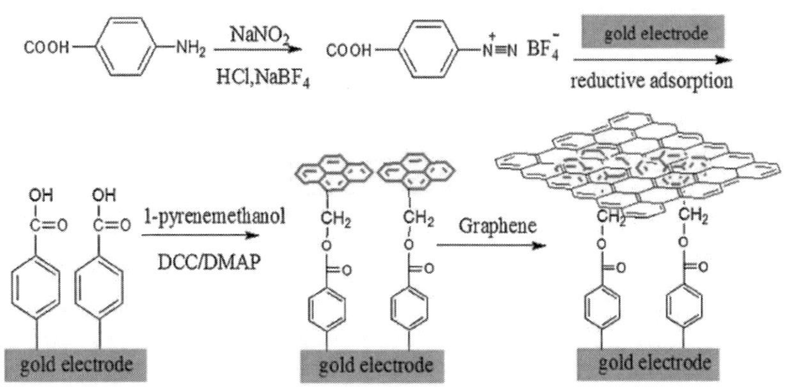

Figure 3. Schematic of fabrication of graphene attached gold electrode using step-by-step surface modifications.

Some researchers developed graphene based Pb^{2+} sensors in the presence of Hg^{2+}.[87,88] Zhao et al.[87] presented the ultra-sensitive and selective sensing Pb^{2+} by stripping voltammetry beyond the selective adsorption of polypyrrole-reduced graphene oxide (PPy-rGO) nanocomposite toward Hg^{2+}. Electrochemical experiments showed that Pb^{2+} could be reliably measured at very low levels (Sensitivity: 0.642 μA nM^{-1}; LOD: 4 pM) on the modified electrode in the presence of Hg^{2+}. Later, Hg^{2+}/GO modified GCE was also fabricated for the detection of Pb^{2+} by the reaction of graphene oxide (GO) with Hg^{2+} first, followed by coating Hg^{2+}/GO dispersion in a Nafion solution on the polished GCE.[88] Compared with the Hg^{2+} immobilized MWCNT graphite electrode, the response signal of Hg^{2+}/GO modified GCE was obviously increased owing to the unique structure of GO. Experimental results showed that the linear calibration curves ranged from 5 ng L^{-1} to 70 ng L^{-1} and 0.1 g L^{-1} to 10 μg L^{-1}, with the detection limit (S/N = 3) 0.13 ng L^{-1} for Pb^{2+}. Hg^{2+}/GO modified GCE were also evaluated by the detection of Pb^{2+} in different water samples and results were satisfying.

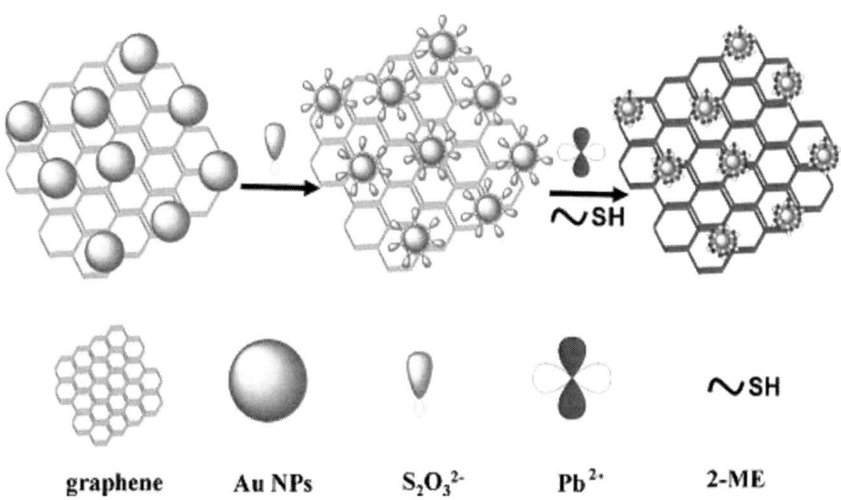

Figure 4. Schematic representation of the sensing mechanism for the detection of Pb^{2+} ions based on accelerated leaching of gold nanoparticles on the surface of graphene.

Graphene based sensor for the determination of Hg^{2+} was first reported by Gong and coworkers.[89] The sensor was based on an AuNPs-graphene hybrid nanocomposite by homogenously distributing monodispersed AuNPs onto the 2D graphene nanosheet matrix. Such a nanostructured composite film greatly facilitated electron-transfer processes and the sensing behavior for Hg^{2+} detection, leading to a remarkably improved sensitivity and selectivity. Later, Chen et al.[90] fabricated an Hg^{2+} ion sensor using thermally reduced graphene oxide (rGO) sheets decorated with thioglycolic acid (TGA) functionalized AuNPs. To verify the selectivity of the rGO/TGA-AuNP hybrid sensor for the detection of Hg^{2+}, Sodium, calcium, zinc, cadmium, iron, copper, and lead ions were also investigated. Results showed that the as-fabricated hybrid sensor had excellent Hg^{2+} ion sensing performance with detection limit of 2.5×10^{-8} M and a response time as fast as a few seconds.

Zhou et al.[91] also established a novel Hg^{2+} sensor based on ionic liquid functionalized graphene oxide (GO-IL) and gold nanoparticles (AuNPs) modified glassy carbon electrode (GCE). The as-fabricated Hg^{2+} sensor, based on the combination of the specific ability of GO-IL (i.e., unique electrical conductivity and enlarged active surface area) and the unique properties of the AuNPs (i.e., extraordinary catalytic activity and good conductivity), was successfully tested in drinking and environmental water samples. The limit of detection was estimated to be 0.03 nM at a signal-to-noise ratio of 3 s. In addition, to verify high selectivity of the as-fabricated Hg^{2+} sensor, a variety of common coexistent ions in water samples were investigated, showing no obvious interferences on the Hg^{2+} detection. An ultrasensitive and highly selective Hg^{2+} sensor through non-covalent modification of an electrochemically reduced GO (ERGO)-based diode with N-[(1-pyrenyl-sulfonamido)-heptyl]-gluconamide (PG) as the modifier was designed and fabricated by Yu and coworkers.[84] The glucose residue of PG had imines and hydroxyl functional groups, which could work as multiple-receptor sites for Hg^{2+} in the medium. The detection limit of this sensor to the analyte was 0.1 nM, which is about 10 times as low as that reported previously.[92] The simple fabrication method and excellent sensor performance may open up the opportunity to explore the modified ERGO sensor as an integratable Hg^{2+} practical detector.

3.3. Organic Molecules Detection

Graphene based chemical sensors have also been developed by incorporating graphene sheets into other materials which enhance the electrochemical properties by the synergic effects for the detection of organic molecules. Kang and coworkers[93] fabricated an electrochemical sensor with graphene-modified glassy carbon electrode (GCE) for ultrasensitive determination of paracetamol in pharmaceutical products. This electrochemical sensor showed an excellent performance for detecting paracetamol with a detection limit of 3.2×10^{-8} M, a reproducibility of 5.2 % relative standard deviation, and a satisfied recovery from 96.4 % to 103.3 %. Such electrocatalytic behavior of graphene was attributed to its unique physical and chemical properties, e.g., subtle electronic characteristics, attractive π–π interaction, and strong adsorptive capability. Fan et al.[94] also reported a sensitive and selective method for detection of bisphenol A (BPA) in water samples utilizing nitrogen-doped graphene sheets (N-GS) and chitosan as electrode modification materials. It indicated that compared with graphene, N-GS had favorable electron transferability and electrocatalytic property, leading to significantly enhanced response signal of BPA. Later, a novel electrochemical sensor based on stacked graphene nanofibers (SGNF) and gold nanoparticles (AuNPs) composite modified glassy carbon electrode (GCE) was constructed for the sensitive detection of BPA.[95] The effective surface areas of the AuNPs/SGNF modified electrode increased for about 1.7-fold compared with that of the bare GCE. In addition, a good linear relationship between the peak current and BPA concentration was also obtained in the range from 0.08 to 250 μM with a detection limit of 3.5×10^{-8} M. The modified electrode showed excellent sensitivity, reproducibility and long-term stability for the determination of BPA and was successfully applied for the determination of BPA in baby bottles with satisfying results. Similarly, an electrochemical sensor was prepared by directly electrodepositing poly(3,4-ethylenedioxy-thiophene) (PEDOT)/graphene oxide (GO) hybrid film on GCE for the simultaneous detection of hydroquinone (HQ) and catechol (CT).[96] The modified GCE showed excellent electrocatalytic activity toward the redox of HQ and CT. The results indicated that under the optimized condition, the response peak currents of the modified electrodes were linear over ranges from 2.5 to 200 μM for HQ and from 2 to 400 μM for CT. The sensor also exhibited good sensitivity with the detection limit of 1.6 μM for both HQ and CT, and good stability. A surface imprinted sensor based on an electrochemical reduced graphene modified carbon electrode was also prepared for indirect detection of 3,3',5,5'-tetrabromobisphenol A (TBBPA) and the detection was successfully implemented by monitoring the peak current of $Fe(CN)6^{3-/4-}$–TBBPA complex.[97] The response currents of the imprinted sensor showed a linear relationship toward the TBBPA concentration range from 0.5 to 4.5 nM with a detection limit of 0.23 nM (S/N = 3). Due to high sensitivity and simple operation, this determination methodology became a promising candidate for the detection of TBBPA in real water samples.

In addition, Ameen et al.[98] studied the modified electrode of polyaniline/graphene (PANI/Gr) composites by in situ electrochemical synthesis of aniline monomer and graphene oxide in hydrochloric acid for the rapid detection of hydrazine through simple current (I)–voltage (V) characteristics. The fabricated hydrazine sensor showed very high sensitivity of ~ 32.54×10^{-5} A cm^{-2} mM^{-1} and high detection limit ~ 15.38 mM. This simple and sensitive sensor had a great promise for the detection of hydrazine. An amine functionalized reduced graphene oxide/carbon nanotube nanocomposite (a-rGO/CNT) was also fabricated for the detection of trinitrotoluene (TNT) utilizing electron rich ethylenediamine as a receptor for binding of electron-deficient trinitrotoluene to form specific charge-transfer Jackson–Meisenheimer (JM) complex.[99] The synergistic effect of rGO/CNT with large electroactive surface area contributed fast charge transfer thereby enhancing the electrochemical signal sensitivity up to 0.01 ppb in standard TNT samples. Thereby, the a-rGO/CNT sensing platform could be an alternate for sensitive detection of hazardous environmental pollutant trinitrotoluene. Liu and coworkers[100] also reported a molecular imprinted polymer (MIP) sensor based on electrochemical reduction of GO for the detection of chlortetracycline (CTC).

Except for the above graphene based sensors, fluorescence resonance energy transfer (FRET) sensor and photoinduced electron transfer sensor were also developed to detect some organic molecules. Huang et al.[101] reported a fast, sensitive, label-free, and general dye-sensor for synthetic organic dyes detection by competitive adsorption based on fluorescein/reduced graphene oxide (Fl/rGO) complex. Results showed that rGO could bind Fl and quench the fluorescence by FRET, while methylene blue (MB) could quickly make the fluorescence recover by displacing Fl from the Fl/rGO complex, which provided a quantitative readout for MB. Because this assay principle was simply based on the reversible interaction (adsorption and desorption) of rGO and FD, so it could generally be applied to the detection of various synthetic organic dyes. Likewise, a fluorescent "turn-on" sensor based on 8-aminoquinoline functionalized graphene oxide for detection of D-glucosamine with a high selectivity and sensitivity through a photoinduced electron transfer (PET) signaling mechanism was also prepared.[102]

4. Graphene Based Biosensors

Biosensor is based on the degree of biological activity unit (such as enzymes, antibodies, antigens, microbes, cells, tissues, nucleic acid) as the biologically sensitive primitives, and captures the reaction between the target molecule and sensitive primitives by various physical, chemical signal converter. The biosensor will then express the reaction with a discrete or continuous electrical signal. Therefore, high conductivity, good biocompatibility and excellent chemical stability make graphene and its derivatives become ideal platforms for biosensors in aqueous environment. In this section, we describe the application of graphene based biosensors for the detection of glucose, DNA, hydrogen peroxide (H_2O_2) and others compounds.

4.1. Glucose Detection and Diabetes Diagnosis

Electrochemical biosensors based on enzymes are suitable for a highly selective, sensitive, and rapid analysis of various biological species in vivo and in vitro. As a result, biosensors based on enzymes are extensively studied. Glucose oxidase (GOD) is often used for the fabrication of most glucose biosensors and bioelectronic devices, such as biofuel cells. It catalyzes the oxidation of glucose to gluconolactone, while itself is reduced and deactivated. In the presence of oxygen or other oxidants the deactiviated GOD can be reactivated following the process as represented in the equation below[103,104]:

$$\mathrm{GOD(FAD)} + \mathrm{glucose} \rightarrow \mathrm{GOD(FADH_2)} + \mathrm{Gluconolactone}$$
$$\mathrm{GOD(FADH_2)} + O_2 \rightarrow \mathrm{GOD(FAD)} + H_2O_2$$

The first glucose sensor was developed by Clark and Lyons from the Cincinnati Children's Hospital in 1962,[105] and the entire field of biosensors can trace its origin to this original glucose enzyme electrode. Clark's first generation glucose biosensor used one or more enzymes for converting electroinactive substrates to electroactive products which the detection limit was very low. However, since the first product emerged, the entire glucose sensor market has grown rapidly with higher sensitivity and larger detection range.

Recently, attaching glucose oxidase (GOD) on functional graphene films was widely applied in fabricating glucose biosensors. Huang et al.[106] demonstrated the use of large-sized CVD grown graphene/glucose oxidase (GOD) films configured as field-effect transistors for real-time glucose biosensing. Glucose oxidase (GOD) was immobilized onto graphene film via a linker molecule (1-pyrenebutanoic acid succinimidyl ester) which on one end firmly attaches to the graphene surface through π–π interaction with a pyrene group and on the other end covalently reacts with the amino group on the enzyme by an amide bond. The detection limit of the graphene based glucose sensor can be low to 0.1 mM. In another research, the GOD was immobilized on reduced graphene oxide (RGO) via an electrochemical approach without any crosslinking agents or modifiers yet.[107] A simple solution phase approach was used to prepare exfoliated graphene oxide (GO), followed by electrochemical reduction to afford RGO–GOD biocomposite. It is notable that the developed biosensor exhibits excellent catalytic activity towards glucose over a wide linear range of 0.1 – 27 mM with a sensitivity of 1.85 μA mM^{-1} cm^2 in mediator-free conditions. Compared with reduced graphene sheets, microporous graphene oxide exhibits good biocompatibility and has potential advantages for graphene-based biosensing application.[108-110] In addition, a highly efficient enzyme electrode was directly obtained via covalent attachment between carboxyl acid groups of graphene oxide sheets and amines of GOD.[109] The graphene oxide based glucose biosensor exhibited a broad linear range, high sensitivity and good reproducibility and storage stability.

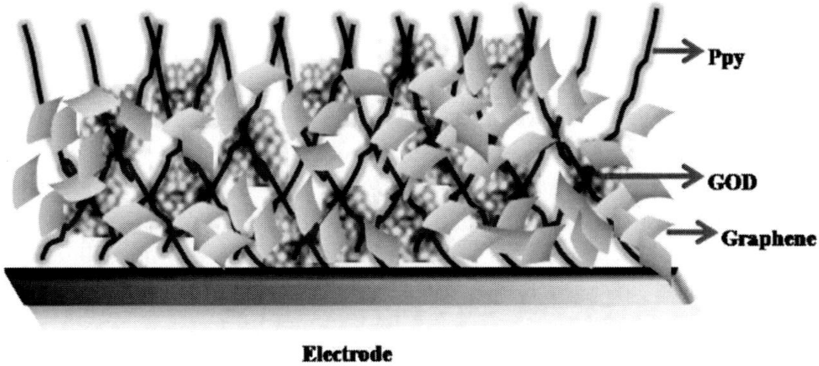

Figure 5. Representation of graphene–GOD entrapped within a porous Ppy matrix.

To protect the enzymes on the electrode, some polymers like polypyrrole, polyethylenimine, polyvinylpyrrolidone and nafion were utilized as shown below. Li and his co-authors[111] described a method in which graphene is covalently conjugated with GOD and immobilized onto the polypyrrole (Ppy) electrochemically modified electrode surface. As shown in Figure 5, the polypyrrole (Ppy) forms a stable and porous matrix that will encapsulate graphene-GOD on its surface. As expected, Ppy-graphene-GOD electrodes exhibited an excellent sensitivity of 3 μM for glucose detection which was higher than that based on other carbon materials.[112,113] Similarly, Shan et al.[114] constructed a polyvinylpyrrolidone-protected graphene/ polyethylenimine-functionalized ionic liquid/GOD electrochemical biosensor with glucose detection linear response up to 14 mM. The polymeric ionic liquid/graphene nanocomposite was also prepared for GOD immobilization for potential application for the fabrication of novel glucose biosensors.[115] The graphene/nafion film was also utilized to encapsulate the GOD for glucose biosensing.[116]

In a recent study by Liu and coworkers the multi-layered GOD enzyme electrodes were fabricated with controlled alternate enzyme and graphene layers,[117] where GOD was first modified with pyrene functionalities in order to be self-assembled onto a graphene basal plane via non-covalent π-π stacking interaction. Via alternate layer-by-layer self-assembly of graphene and pyrene functionalized GOD, mono- and multi-layered enzyme electrodes with controlled biocatalytical activity could be easily fabricated. (Figure 6 A) The biocatalytical activity of the as-prepared enzyme electrodes increased with increasing graphene and GOD layers and came into saturation when the layers reached four. (Figure 6 B) Such multi-layered enzyme electrodes with controlled nanostructure exhibited reliable application in human serum samples analysis with higher detection sensitivity, good stability and repeatability. Some special nanostructures such as graphene platelet–glucose oxidase (GP–GOD) nanostructures [118] have been prepared through self-assembly of GOD and chitosan functionalized graphene sheets by electrostatic attraction in aqueous solution. It was found that the resulting biosensor exhibits good response to glucose. The linear detection range is estimated to be from 2 to 22 mM ($r = 0.9987$), and the detection limit is estimated to be 20 μM at a signal-to-noise (S/N) ratio of 3.

4.2. DNA Detection

The electrochemical detection of DNA has also attracted a significant amount of attention. The need for a DNA biosensor with low cost, high sensitivity and good selectivity is becoming a very important issue in clinical diagnosis. Several works about the biosensors based on the graphene have been published and these results showed that the graphene-based DNA biosensor presented excellent sensitivity.[119] Pumera et al. developed a graphene platform combined hairpin-shaped DNA probes for the detection of DNA hybridization and polymorphism via electrochemical impedance spectroscopy as the detection technique.[120] Moreover, through comparing the performance of three different graphene platforms, they found the most sensitive graphene based biosensor for single nucleotide polymorphism detection. (Figure 7)

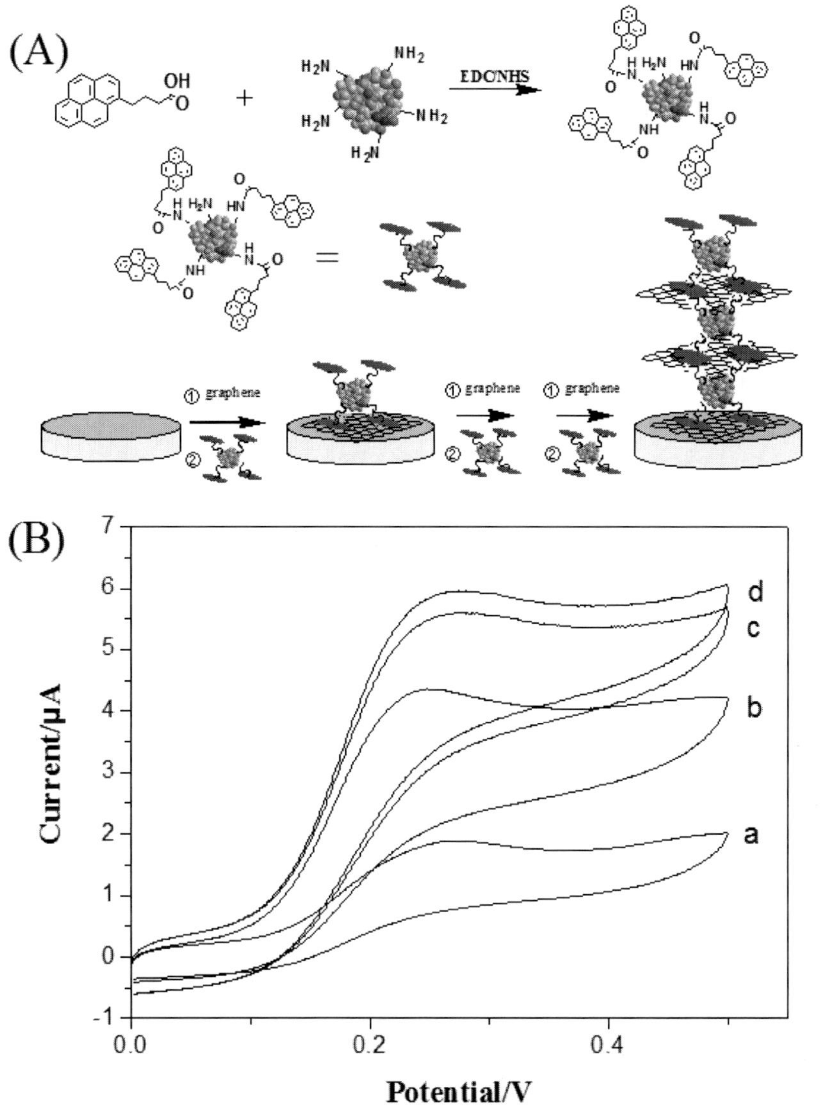

Figure 6. (A) Schematic diagram of the modification of GOD with pyrene and the subsequent fabrication of mono- and multi-layered enzyme electrodes. (B) Cyclic voltammograms of the enzyme electrodes fabricated with mono- and multi-layered graphene and pyrene functionalized GOD: (a) 1 layer, (b) 2 layers, (c) 3 layers, and (d) 4 layers. Data were recorded in 80 mM glucose solution in 0.1 M pH 7.4 phosphate buffer, at room temperature and potential scan rate of 10 mV s^{-1}.

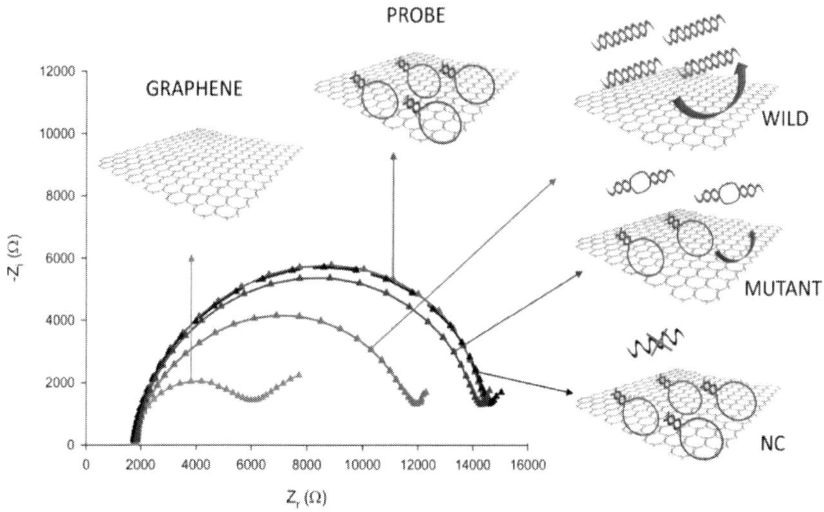

Figure 7. Schematic of the protocol and Nyquist plots, $-Z_i$ vs Z_r, of the graphene surface (gray), hpDNA (blue), complementary target (red), 1-mismatch target (green), and negative control with a noncomplementary sequence (black) (concentration of the DNA probes, 1×10^{-5} M; concentration of the DNA target, 3×10^{-8} M). All measurements were performed in 0.1 M PBS buffer solution containing 10 mM $K_3[Fe(CN)_6]/K_4[Fe(CN)_6]$.

Huang and his co-workers developed another DNA biosensor utilized the oxidized graphene and polyaniline nanowires (PANIws) modified glassy carbon electrode.[121] The resulting graphene/PANIw layers exhibited good differential pulse voltammetry (DPV) current response and good electron transfer activity which might be attributed to the effect of graphene and PANIw. After immobilization of the DNA probe, the resulted graphene/PANIw/DNA biosensor exhibited a fast amperometric response, high sensitivity (detection limit is 3.25×10^{-13} mol L^{-1}) and good storage stability for sensitive and selective DNA detection. Moreover, a chemically modified graphene paste electrode was prepared by Huang's group, they incorporated appropriate amounts of graphene in a paste mixture, followed by electrodepositing Prussian blue (PB) and coating chitosan on the electrode surface.[122] PB could express good electrochemical properties under the protection of chitosan film, so the electrode was able to bind single-stranded DNA (ssDNA), and gave a better voltammetric response for complement DNA than did ordinary carbon paste electrodes. The technique for establishing DNA biosensor with high selectivity can offer a new approach for developing novel types of highly sensitive and stable electrochemical biosensors. Some other research about the graphene based DNA biosensor with high selectivity which could discriminate completely complementary target sequence has been published in recent years.[123-128]

4.3. Hydrogen Peroxide Detection

Accurate determination of hydrogen peroxide (H_2O_2) is very important in chemistry, biology, clinical control and environmental protection. The electrochemical methods are more convenient for the H_2O_2 detection than other conventional methods. Horse radish peroxidase (HRP) can be used to decompose the H_2O_2 into H_2O and O_2, so when the HRP attached on the electrode, we can indirectly detect the concentration of H_2O_2 according to the above redox reaction.[129,130] Gu et al.[131] have reported an ultrasensitive hydrogen peroxide biosensor, which was fabricated by coating graphene–gold nanocomposite, CdTe–CdS core–shell quantum dots, gold nanoparticles and HRP in sequence on the surface of gold electrode. The biosensor displayed very high sensitivity, low detection limit (S/N = 3) (3.2×10^{-11} M) and good long-term stability (20 weeks). Lu and other researchers[132] also introduced the direct electrochemistry hydrogen peroxide biosensor via single-layer graphene nanoplatelet and HRP hybrid for H_2O_2 analysis. Through the layer-by-layer self-assembly, the hierarchical nanostructures of HRP and sodium dodecyl benzene sulphonate (SDBS) functionalized graphene sheets (GSs) was constructed via electrostatic attraction in aqueous solution.[133] The so-called HRP–GSs electrodes displayed high electrocatalytic activity to H_2O_2 with high sensitivity, wide linear range, low detection limit, and fast amperometric response which can be contributed to graphene's superb electron transport efficiency as well as high HRP loading and synergistic catalytic performance. In addition, many non-enzyme hydrogen peroxide biosensors without any enzyme were also reported.[25,134-138]

4.4. Immunosensors

Immunosensors are one of the most important analytical sensors for sensitive and selective detection of proteins; they have found wide applications in clinical diagnosis, biomedical research, food quality control and environmental monitoring.[139-143] In recent years, the research on enhancing immunosensors' performance and optimizing its examining method are increasingly developed. Graphene-based immunosensors can be achieved by immobilizing the biomolecules, such as antibodies or designing technology of sensing interface to capture the corresponding antigens in the real samples.

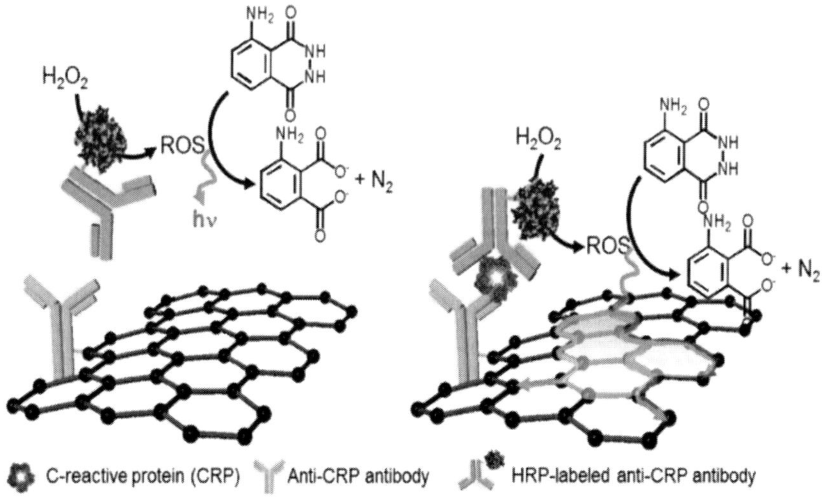

Figure 8. Schematic illustration of a graphene-based chemiluminescence resonance energy transfer platform for the detection of C-reactive protein (CRP). Anti-CRP antibody-conjugated graphene and luminol that is excited by horseradish peroxidase (HRP)-catalyzed oxidation are used as an acceptor and a donor, respectively.

Graphene-based chemiluminescence resonance energy transfer (CRET) platform can also be produced for homogeneous immunoassay of C-reactive protein (CRP) via a luminol/hydrogen peroxide chemiluminescence (CL) reaction catalyzed by horseradish peroxidase.[144] As shown in Figure 8, graphene played a key role as an energy acceptor, which was more efficient than graphene oxide, while luminol served as a donor to graphene, triggering the CRET phenomenon between luminol and graphene. The graphene-based CRET platform was successfully applied to the detection of CRP in human serum samples in the range observed during acute inflammatory stress with a detection limit above 1.6 ng mL^{-1}. Mao et al.[145] constructed an electrochemical immunosensor for prostate specific antigen (PSA) detection using a nanocomposite film of graphene sheets–methylene blue–chitosan (GS-MB-CS) as electrode material. The nanocomposite film showed high binding affinity to the electrode and was used to immobilize the antibody of PSA with a low limit of detection (13 pg mL^{-1}) and a high selectivity. An electrochemical sensor was fabricated for label-free cancer cell detection using the first clinical trial II used aptamer AS1411 and functionalized graphene.[140] By taking advantages of AS1411 high binding affinity and specificity to the overexpressed nucleolin on the cancer cell surface, the electrochemical aptasensor can distinguish cancer cells and normal ones and detect as low as one thousand cells. In addition to cancer cells, the graphene-based immunosensors have been also applied for the detection of bacterium,[119,142,146] pathogenic virus,[147] and others.[148]

4.5. NADH Detection

The NADH oxidase, a plasma membrane-bound protein, is sensitive to activation by growth factors, including epidermal growth factor and insulin.[149] Shan et al.[150] showed that low-potential NADH detection and biosensing for ethanol were achieved at an ionic liquid-functionalized graphene (IL-graphene) modified electrode. The IL-graphene-based sensor for NADH exhibited very good analytical performance with low cost, convenient preparation, and sensitive, rapid, reproducible detection. the electrochemical behavior of NADH on reduced graphene sheet films (rGSF) was also studied by Tang et al.[151]. It showed increased excellent electrocatalytic activity compared with the bare GC electrodes.

4.6. Organophosphates (OPs) and Nitric Oxides (NOs) Detection

Nitric oxides, organophosphates are new poisonous pollutants that require lot of effort to remove, so it is important to monitor the pollutant timely. Wang et al.[152] developed a rapid and sensitive OPs amperometric biosensor via immobilizing the acetylcholinesterase (AChE) on CdS–decorated graphene (CdS–G) nanocomposite. The as-prepared biosensor exhibited high affinity to acetylthiocholine (ATCl) with a Michaelis–Menten constant (K_m) value of 0.24 mM and a rapid inhibition time of 2 min. It is more efficient than that of TiO_2-decorated graphene (TiO_2-G) nanohybrid amperometric biosensor for OPs with a Michaelis–Menten constant (K_m) value of 0.22 mM, and rapid inhibition time (3 min).[153] Choi and co-authors[154] prepared a free-standing flexible conductive reduced graphene oxide/nafion (RGON) hybrid films by a solution chemistry that utilizes self-assembly and directional convective-assembly to analyze the Ops with a sensitivity of 10.7 nA M^{-1}, detection limit of 1.37 × 10^{-7} M, and response time of < 3 s. These strategies described above provided insight into the design and fabrication of graphene and hybrid nanomaterials for practical analytical device applications.

A sensitive nitrite biosensor[155] was developed through a sensing platform consisted of single-layer graphene nanoplatelet (SLGnP)–protein composite film. SLGnP composite film showed virtues of excellent biocompatibility, conductivity, high sensitivity to the local perturbations, which could be contributed to the biocompatible microenvironment for protein immobilization and the suitable electron transfer distance between heme protein and electrode. The nitromethane (CH_3NO_2) amperometric biosensor[156] based on immobilization of graphene, chitosan, hemoglobin and room temperature ionic liquid (IL) on a glassy carbon electrode (GCE) was developed with the apparent Michaelis–Menten constant down to 0.16 mM, fast amperometric response (< 5 s), low detection limit (6.0×10^{-10} M), and excellent long-time storage stability for the determination of CH_3NO_2. A nitric oxide biosensor based on chitosan-dispersed graphene nano-flakes and cytochrome c modified glassy carbon electrode was also developed.[157] Cytochrome c on the surface of electrode maintained its bioactivity and showed an enzyme-like activity for the reduction of nitric oxide, displaying a potential application for the fabrication of novel biosensors to sense nitric oxide. This research will enlarge the applications of graphene-based materials in biosensor field.

4.7. Other Molecules and Substance Detection

The availability of 2D graphene will open up possibilities for designing and preparing graphene-oriented electrodes for a wide range of electrochemical sensing and biosensing applications. For example, to detect the heavy metal ions, Liu et al. [158] designed a novel and promising "turn-on" fluorescent Cu^{2+} biosensor based on graphene–DNAzyme catalytic beacon. Compared with common DNAzyme-based sensors, the presented graphene-based catalytic beacon greatly improved the signal-to-background ratio, hence increased the sensitivity (LOD = 0.365 nM). Furthermore, the controllable DNA cleavage reaction provided an original and alternative internal method to regulate the interaction between graphene and DNA relative to the previous external sequence-specific hybridization-dependent regulation, which will open new opportunities for nucleic studies and sensing applications in the future. A similar graphene DNAzyme based biosensor for amplified fluorescence "turn-on" detection of Pb^{2+} was also proposed.[159] It was found that the biosensor exhibited a high sensitivity toward the target with a detect ion limit of 300 pM for Pb^{2+}, which is lower than previously reported for catalytic beacons.

A new type of chemically modified graphene, EDTA-modified reduced graphene (EDTA-RG), was synthesized by silanization of graphene with N-(trimethoxysilylpropyl) ethylenediamine triacetic acid (EDTA-silane). It was found that the EDTA-RG-Nafion modified electrode can be used to detect dopamine and ascorbic acid.[160] It is also very important application of graphene for analysis l-cysteine,[161] cholesterol,[138] hemoglobin,[162] thrombin,[163], aromatic compounds (such as phenol,[164] and catechol[165]) and other compounds.[166,167]

CONCLUSION

Graphene, as the shining star among nanomaterials and a hotspot among researchers all over the world, has drawn considerable interests in many fields. Its unique structure contributes to its exceptional chemical and physical properties, which leads to the wide applications in energy and environmental fields such as solar cells, supercapacitors, electrode devices, absorption and sensors. In this chapter, the recent development of graphene based chemical and biological sensors for the detection of gas, heavy metal ions, biological molecules, protein and bacterias was briefly reviewed. The amazing properties of graphene did certainly lead to many future advances in science and technology, and provide convenience to the public. However, graphene as a new material still faces many challenges ranging from synthesis and extensive use in various fields when employed for practical application. In the case of synthesis, some fundamental approaches are provided including mechanical and matrix assisted exfoliations, thermal deposition, CVD, unzipping CNT, and reduction of graphite oxide. Each of these preparation methods has some advantages for specific applications, but also some drawbacks. In the future, graphene will still be the focus of the scientific community and more efficient and novel methods of preparing high quality, large-sized and defect-free graphene are expected to be developed. Moreover, with the joint efforts by scientists in physics, chemistry and biology, more outstanding properties of graphene will be discovered and more efficient chemo- and bio-sensors will be developed.

ACKNOWLEDGMENTS

J. Liu thanks the NSF of China (51173087), NSF of Shandong (ZR2011EMM001) and Taishan Scholars Program for financial support. S. Feng thanks the NSF of China (21274080 and 20874057) Program for financial support.

REFERENCES

[1] Novoselov, K.; A. Geim. (2004). Electric field effect in atomically thin carbon films. *Science*, 306, 666-669.

[2] Shukla, A.; R. Kumar. (2009). Graphene made easy: high quality, large-area samples. *Solid State Communications,* 149, 718-721.

[3] Hernandez, Y.; V. Nicolosi. (2008). High-yield production of graphene by liquid-phase exfoliation of graphite. *Nature Nanotechnology*, 3, 563-568.

[4] Lotya, M.; Y. Hernandez. (2009). Liquid phase production of graphene by exfoliation of graphite in surfactant/water solutions. *Journal of the American Chemical Society*, 131, 3611-3620.

[5] Berger, C.; Z. Song. (2004). Ultrathin epitaxial graphite: 2D electron gas properties and a route toward graphene-based nanoelectronics. *The Journal of Physical Chemistry B,* 108, 19912-19916.

[6] Lin, Y.M.; C. Dimitrakopoulos. (2010). 100-GHz transistors from wafer-scale epitaxial graphene. *Science*, 327, 662-662.

[7] Somani, P.R., S.P. Somani. (2006). Planer nano-graphenes from camphor by CVD. *Chemical Physics Letters,* 430, 56-59.

[8] Bae, S.; H. Kim. (2010). Roll-to-roll production of 30-inch graphene films for transparent electrodes. *Nature Nanotechnology*, 5, 574-578.

[9] Cano-Márquez, A.G.; F.J. Rodríguez-Macías. (2009). Ex-MWNTs: Graphene sheets and ribbons produced by lithium intercalation and exfoliation of carbon nanotubes. *Nano Letters,* 9, 1527-1533.

[10] Jiao, L.; L. Zhang. (2009). Narrow graphene nanoribbons from carbon nanotubes. *Nature*, 458, 877-880.

[11] Stankovich, S.; D.A. Dikin. (2007). Synthesis of graphene-based nanosheets via chemical reduction of exfoliated graphite oxide. *Carbon*, 45, 1558-1565.

[12] Shin, H.J.; K.K. Kim. (2009). Efficient reduction of graphite oxide by sodium borohydride and its effect on electrical conductance. *Advanced Functional Materials*, 19, 1987-1992.
[13] Su, C.-Y.; A.-Y. Lu. (2011). High-quality thin graphene films from fast electrochemical exfoliation. *ACS Nano,* 5, 2332-2339.
[14] Yen, M.-Y.; C.-K. Hsieh. (2012). Metal-free, nitrogen-doped graphene used as a novel catalyst for dye-sensitized solar cell counter electrodes. *RSC Advances*, 2, 2725-2728.
[15] Biswas, S.; L.T. Drzal. (2010). Multilayered nanoarchitecture of graphene nanosheets and polypyrrole nanowires for high performance supercapacitor electrodes. *Chemistry of Materials,* 22, 5667-5671.
[16] Wang, D.; R. Kou. (2010). Ternary self-assembly of ordered metal oxide− graphene nanocomposites for electrochemical energy storage. *ACS Nano,* 4, 1587-1595.
[17] Dreyer, D.R.; S. Park. (2009). The chemistry of graphene oxide. *Chemical Society Reviews,* 39, 228-240.
[18] Leenaerts, O., B. Partoens. (2008). Adsorption of H_2O, NH_3, CO, NO_2, and NO on graphene: a first-principles study. *Physical Review B,* 77, 125416-125422.
[19] Zhang, N.; H. Qiu. (2011). Fabrication of highly porous biodegradable monoliths strengthened by graphene oxide and their adsorption of metal ions. *Carbon*, 49, 827-837.
[20] Sun, Z.; J. Masa. (2012). Highly concentrated aqueous dispersions of graphene exfoliated by sodium taurodeoxycholate: dispersion behavior and potential application as a catalyst support for the oxygen eduction reaction. *Chemistry-A European Journal,* 18, 6972-6978.
[21] Xin, Y.; J. Liu. (2012). Preparation and electrochemical characterization of nitrogen doped graphene by microwave as supporting materials for fuel cell catalysts. *Electrochimica Acta,* 60, 354-358.
[22] Zhou, K.; Y. Zhu. (2011). Preparation of graphene-TiO_2 composites with enhanced photocatalytic activity. *New Journal of Chemistry,* 35, 353-359.
[23] Blake, P.; P.D. Brimicombe. (2008). Graphene-based liquid crystal device. *Nano Letters,* 8, 1704-1708.
[24] Li, X.; G. Zhang. (2008). Highly conducting graphene sheets and Langmuir lodgett films. *Nature Nanotechnology*, 3, 538-542.
[25] Choucair, M., P. Thordarson. (2008). Gram-scale production of graphene based on solvothermal synthesis and sonication. *Nature Nanotechnology,* 4, 30-33.

[26] Jang, J.H.; D. Rangappa. (2010). Direct preparation of 1-PSA modified graphene nanosheets by supercritical fluidic exfoliation and its electrochemical properties. *Journal of Materials Chemistry,* 21, 3462-3466.
[27] Liu, F., J.Y. Choi. (2010). DNA mediated water-dispersible graphene fabrication and gold nanoparticle-graphene hybrid. *Chemical Communications,* 46, 2844-2846.
[28] Lee, D.W., T. Kim. (2011). An amphiphilic pyrene sheet for selective functionalization of graphene. *Chemical Communications,* 47, 8259-8261.
[29] Liu, Z.; J. Liu. (2012). Preparation of graphene/polymer composites by direct exfoliation of graphite in functionalised block copolymer matrix. *Carbon*, 51,148-155.
[30] Yang, Q.; X. Pan. (2010). Fabrication of high-concentration and stable aqueous suspensions of graphene nanosheets by noncovalent functionalization with lignin and cellulose derivatives. *The Journal of Physical Chemistry C*, 114, 3811-3816.
[31] Rollings, E.; G.H. Gweon. (2006). Synthesis and characterization of atomically thin graphite films on a silicon carbide substrate. *Journal of Physics and Chemistry of Solids,* 67, 2172-2177.
[32] Emtsev, K.V.; A. Bostwick. (2009). Towards wafer-size graphene layers by atmospheric pressure graphitization of silicon carbide. *Nature Materials*, 8, 203-207.
[33] Juang, Z.Y.; C.Y. Wu. (2009). Synthesis of graphene on silicon carbide substrates at low temperature. *Carbon*, 47, 2026-2031.
[34] Sutter, P.W., J.I. Flege. (2008). Epitaxial graphene on ruthenium. *Nature Materials,* 7, 406-411.
[35] Vazquez de Parga, A.; F. Calleja. (2008). Periodically rippled graphene: growth and spatially resolved electronic structure. *Physical Review Letters,* 100, 56807-56812.
[36] De Arco, L.G.; Y. Zhang. (2009). Synthesis, transfer, and devices of single-and few-layer graphene by chemical vapor deposition. *Nanotechnology*, 8, 135-138.
[37] Reina, A.; X. Jia. (2008). Large area, few-layer graphene films on arbitrary substrates by chemical vapor deposition. *Nano Letters,* 9, 30-35.
[38] Kim, K.S.; Y. Zhao. (2009). Large-scale pattern growth of graphene films for stretchable transparent electrodes. *Nature*, 457, 706-710.

[39] Yu, Q.; J. Lian. (2008). Graphene segregated on Ni surfaces and transferred to insulators. *Applied Physics Letters,* 93, 113103-113106.
[40] Li, X.; W. Cai. (2009). Large-area synthesis of high-quality and uniform graphene films on copper foils. *Science*, 324, 1312-1314.
[41] Cai, W.; A.L. Moore. (2010). Thermal transport in suspended and supported monolayer graphene grown by chemical vapor deposition. *Nano Letters,* 10, 1645-1651.
[42] Kosynkin, D.V.; A.L. Higginbotham. (2009). Longitudinal unzipping of carbon nanotubes to form graphene nanoribbons. *Nature*, 458, 872-876.
[43] Hummers Jr, W.S.; R.E. Offeman. (1958). Preparation of graphitic oxide. *Journal of the American Chemical Society*, 80, 1339-1339.
[44] Dikin, D.A.; S. Stankovich. (2007). Preparation and characterization of graphene oxide paper. *Nature*, 448, 457-460.
[45] Gao, W.; L.B. Alemany. (2009). New insights into the structure and reduction of graphite oxide. *Nature Chemistry,* 1, 403-408.
[46] Pei, S.; J. Zhao. (2010). Direct reduction of graphene oxide films into highly conductive and flexible graphene films by hydrohalic acids. *Carbon*, 48, 4466-4474.
[47] Bourlinos, A.B.; D. Gournis. (2003). Graphite oxide: chemical reduction to graphite and surface modification with primary aliphatic amines and amino acids. *Langmuir*, 19, 6050-6055.
[48] Xiao, P.; M. Xiao. (2000). Direct synthesis of a polyaniline-intercalated graphite oxide nanocomposite. *Carbon*, 38, 623-641.
[49] Gao, J.; F. Liu. (2010). Environment-friendly method to produce graphene that employs vitamin C and amino acid. *Chemistry of Materials,* 22, 2213-2218.
[50] Glover, A.J.; M. Cai. (2011). In situ reduction of graphene oxide in polymers. *Macromolecules*, 44, 9821-9829.
[51] Chen, W., L. Yan. (2010). Chemical reduction of graphene oxide to graphene by sulfur-containing compounds. *The Journal of Physical Chemistry C*, 114, 19885-19890.
[52] Fan, X.; W. Peng. (2008). Deoxygenation of exfoliated graphite oxide under alkaline conditions: a green route to graphene preparation. *Advanced Materials*, 20, 4490-4493.
[53] Zangmeister, C.D. (2010). Preparation and evaluation of graphite oxide reduced at 220 C. *Chemistry of Materials,* 22, 5625-5629.
[54] Zhu, Y.; S. Murali. (2010). Microwave assisted exfoliation and reduction of graphite oxide for ultracapacitors. *Carbon*, 48, 2118-2122.

[55] Williams, G., B. Seger. (2008). TiO_2-graphene nanocomposites. UV-assisted photocatalytic reduction of graphene oxide. *ACS Nano,* 2, 1487-1491.

[56] Kauffman, D.R.; A. Star. (2010). Graphene versus carbon nanotubes for chemical sensor and fuel cell applications. *Analyst*, 135, 2790-2797.

[57] Schedin, F.; A. Geim. (2007). Detection of individual gas molecules adsorbed on graphene. *Nature materials,* 6, 652-655.

[58] Jeong, H.Y.; D.-S. Lee. (2010). Flexible room-temperature NO_2 gas sensors based on carbon nanotubes/reduced graphene hybrid films. *Applied Physics Letters,* 96, 213105-213108.

[59] Deng, S.; V. Tjoa. (2012). Reduced graphene oxide conjugated Cu_2O nanowire mesocrystals for high-performance NO_2 gas sensor. *Journal of the American Chemical Society*, 134, 4905-4917.

[60] Zhou, L.; F. Shen. (2013). Stable Cu_2O nanocrystals grown on functionalized graphene sheets and room temperature H_2S gas sensing with ultrahigh sensitivity. *Nanoscale*, 5, 1564-1569.

[61] Chung, M.G.; D.H. Kim. (2012). Highly sensitive NO_2 gas sensor based on ozone treated graphene. *Sensors and Actuators B: Chemical*, 166, 172-176.

[62] Dua, V.; S.P. Surwade. (2010). All-organic vapor sensor using Inkjet-printed reduced graphene oxide. *Angewandte Chemie International Edition,* 49, 2154-2157.

[63] Fowler, J.D.; M.J. Allen. (2009). Practical chemical sensors from chemically derived graphene. *Acs Nano,* 3, 301-306.

[64] Yuan, W.; A. Liu. (2012). High-performance NO_2 sensors based on chemically modified graphene. *Advanced Materials,* 25, 766-771.

[65] Lu, G.; K. Yu. (2011). Ultrafast room temperature NH_3 sensing with positively gated reduced graphene oxide field-effect transistors. *Chemical Communication,* 47, 7761-7763.

[66] Yu, K.; P. Wang. (2011). Patterning vertically oriented graphene sheets for nanodevice applications. *The Journal of Physical Chemistry Letters*, 2, 537-542.

[67] Huang, B.; Z. Li. (2008). Adsorption of gas molecules on graphene nanoribbons and its implication for nanoscale molecule sensor. *The Journal of Physical Chemistry C,* 112, 13442-13446.

[68] Shafiei, M.; R. Arsat. (2009). Pt/graphene nano-sheet based hydrogen gas sensor. in Sensors.. *IEEE*, 295-298.

[69] Johnson, J.L.; A. Behnam. (2010). Hydrogen sensing using Pd-functionalized multi-layer graphene nanoribbon networks. *Advanced Materials,* 22, 4877-4880.
[70] Wu, W.; Z. Liu. (2010). Wafer-scale synthesis of graphene by chemical vapor deposition and its application in hydrogen sensing. *Sensors and Actuators B: Chemical,* 150, 296-300.
[71] Yoon, H.J.; D.H. Jun. (2011). Carbon dioxide gas sensor using a graphene sheet. *Sensors and Actuators B: Chemical*, 157, 310-313.
[72] Li, W.; X. Geng. (2011). Reduced graphene oxide electrically contacted graphene sensor for highly sensitive nitric oxide detection. *Acs Nano,* 5, 6955-6961.
[73] Chen, Z.; W. Ren. (2011). Three-dimensional flexible and conductive interconnected graphene networks grown by chemical vapour deposition. *Nature materials*, 10, 424-428.
[74] Jiang, Z.; J. Wang. (2011). A highly efficient chemical sensor material for ethanol: Al_2O_3/Graphene nanocomposites fabricated from graphene oxide. *Chemical Communications,* 47, 6350-6352.
[75] Lipatov, A.; A. Varezhnikov. (2013). Highly selective gas sensor arrays based on thermally reduced graphene oxide. *Nanoscale*, 5, 5426-5434.
[76] Liu, F.; H.D. Ha. (2013). Photoluminescent graphene oxide microarray for multiplex heavy metal ion analysis. *Small*, 9, 3410-3414.
[77] Li, J.; S. Guo. (2009). High-sensitivity determination of lead and cadmium based on the Nafion-graphene composite film. *Analytica chimica acta,* 649, 196-201.
[78] Li, J.; S. Guo. (2009). Nafion–graphene nanocomposite film as enhanced sensing platform for ultrasensitive determination of cadmium. *Electrochemistry Communications*, 11, 1085-1088.
[79] Lien, C.-H.; K.-H. Chang. (2013). Optimizing bismuth-modified graphene-carbon nanotube composite-coated screen printed electrode for lead-ion sensing through the experimental design strategy. *Journal of The Electrochemical Society,* 160, 107-112.
[80] Lien, C.-H.; K.-H. Chang. (2012). Green electrode for Pb^{2+} sensing based on the Nafion-graphene/CNT composite. in Sensors. *IEEE*, 12, 1-3.
[81] Fu, X.; T. Lou. (2012). "Turn-on" fluorescence detection of lead Ions based on accelerated leaching of gold nanoparticles on the surface of graphene. *ACS Applied Materials & Interfaces,* 4, 1080-1086.

[82] Kong, N.; J. Liu. (2013). Graphene modified gold electrode via π–π stacking interaction for analysis of Cu^{2+} and Pb^{2+}. *Sensors and Actuators B: Chemical*, 178, 426-433.

[83] Lu, A.-H.; J.-T. Zhang. (2013). Graphene modified carbon nanosheets for electrochemical detection of Pb (II) in water. *Journal of Materials Chemistry A*, 1-8

[84] Yu, C.; Y. Guo. (2013). Ultrasensitive and selective sensing of heavy metal ions with modified graphene. *Chemical Communications,* 49,6492-6494.

[85] Wonsawat, W.; S. Chuanuwatanakul. (2012). Graphene-carbon paste electrode for cadmium and lead ion monitoring in a flow-based system. Ta*l*anta, 100, 282-289.

[86] Lu, Z.; S. Yang. (2013). A glassy carbon electrode modified with graphene, gold nanoparticles and chitosan for ultrasensitive determination of lead (II). *Microchimica Acta*, 1-8.

[87] Zhao, Z.-Q.; X. Chen. (2012). Beyond the selective adsorption of polypyrrole-reduced graphene oxide nanocomposite toward Hg^{2+}: Ultra-sensitive and-selective sensing Pb^{2+} by stripping voltammetry. *Electrochemistry Communications,* 23, 21-24.

[88] Zhao, Q.; Y. Chai. (2013). Square wave anodic stripping voltammetry determination of lead based on the Hg (II) immobilized graphene oxide composite film as an enhanced sensing platform. *Sensors and Actuators B: Chemical,* 178, 379-384.

[89] Gong, J.; T. Zhou. (2010). Monodispersed Au nanoparticles decorated graphene as an enhanced sensing platform for ultrasensitive stripping voltammetric detection of mercury (II). *Sensors and Actuators B: Chemical,* 150, 491-497.

[90] Chen, K.; G. Lu. (2012). Hg (II) ion detection using thermally reduced graphene oxide decorated with functionalized gold nanoparticles. *Analytical chemistry*, 84, 4057-4062.

[91] Zhou, N.; J. Li. (2013). A functional graphene oxide-ionic liquid composites–gold nanoparticle sensing platform for ultrasensitive electrochemical detection of Hg_{2+}. *Analyst*, 138, 1091-1097.

[92] Sudibya, H.G.; Q. He. (2011). Electrical detection of metal ions using field-effect transistors based on micropatterned reduced graphene oxide films. *Acs Nano,* 5, 1990-1994.

[93] Kang, X.; J. Wang. (2010). A graphene-based electrochemical sensor for sensitive detection of paracetamol. *Talanta*, 81, 754-759.

[94] Fan, H.; Y. Li. (2012). Electrochemical bisphenol a sensor based on N-doped graphene sheets. *Analytica chimica acta,* 711, 24-28.
[95] Niu, X.; W. Yang. (2013). A novel electrochemical sensor of bisphenol A based on stacked graphene nanofibers/gold nanoparticles composite modified glassy carbon electrode. *Electrochimica Acta,* 98, 167-175.
[96] Si, W.; W. Lei. (2012). Electrodeposition of graphene oxide doped poly (3, 4-ethylenedioxythiophene) film and its electrochemical sensing of catechol and hydroquinone. *Electrochimica Acta*, 85, 295-301.
[97] Chen, H.-J.; Z.-H. Zhang. (2013). Molecularly imprinted electrochemical sensor based on a reduced graphene modified carbon electrode for tetrabromobisphenol A detection. *Analyst*, 138, 2769-2776.
[98] Ameen, S., M.S. Akhtar. (2012). Hydrazine chemical sensing by modified electrode based on in-situ electrochemically synthesized polyaniline/ graphene composite thin film. *Sensors and Actuators B: Chemical,*173,177-183.
[99] Sablok, K.; V. Bhalla. (2013). Amine functionalized graphene oxide/CNT nanocomposite for ultrasensitive electrochemical detection of trinitrotoluene. *Journal of hazardous materials*, 248, 322-328.
[100] Liu, Y.; L. Zhu. (2013). Fabrication of molecular imprinted polymer sensor for chlortetracycline based on controlled electrochemical reduction of graphene oxide. *Sensors and Actuators B: Chemical*, 185, 438-444.
[101] Huang, S.T.; Y. Shi. (2012). Fast and sensitive dye-sensor based on fluorescein/reduced graphene oxide complex. *Analyst*, 137, 2593-2599.
[102] Cheng, R.; Y. Liu. (2012). Optical turn-on sensor based on graphene oxide for selective detection of D-glucosamine. *Analytical chemistry,* 84, 5641-5644.
[103] Yang, M.H.; B.G. Choi. (2010). Development of a glucose biosensor using advanced electrode modified by nanohybrid composing chemically modified graphene and ionic liquid. *Electroanalysis*, 22, 1223-1228.
[104] Wang, J. (2008). Electrochemical glucose biosensors. *Chemical reviews,* 108, 814-825.
[105] Clark, L.C.; C. Lyons. (1962). Electrode systems for continuous monitoring in cardiovascular surgery. *Annals of the New York Academy of Sciences,* 102, 29-45.
[106] Huang, Y.; X. Dong. (2010). Nanoelectronic biosensors based on CVD grown graphene. *Nanoscale*, 2, 1485-1488.

[107] Unnikrishnan, B., S. Palanisamy. (2012). A simple electrochemical approach to fabricate a glucose biosensor based on graphene-glucose oxidase biocomposite. *Biosensors and Bioelectronics*, 39, 70-75.

[108] Qin, Y.; Y. Kong. (2012). In situ synthesis of highly loaded and ultrafine Pd nanoparticles-decorated graphene oxide for glucose biosensor application. *Journal of Materials Chemistry,* 22, 24821-24826.

[109] Liu, Y.; D. Yu. (2010). Biocompatible graphene oxide-based glucose biosensors. *Langmuir*, 26, 6158-6160.

[110] Song, Y.; K. Qu. (2010). Graphene oxide: intrinsic peroxidase catalytic activity and its application to glucose detection. *Advanced Materials,* 22, 2206-2210.

[111] Alwarappan, S.; C. Liu. (2010). Enzyme-doped graphene nanosheets for enhanced glucose biosensing. *The Journal of Physical Chemistry C,* 114, 12920-12924.

[112] Lee, D.; J. Lee. (2005). Simple fabrication of a highly sensitive and fast glucose biosensor using enzymes immobilized in mesocellular carbon foam. *Advanced Materials,* 17, 2828-2833.

[113] Lin, Y.; F. Lu. (2004). Glucose biosensors based on carbon nanotube nanoelectrode ensembles. *Nano Letters*, 4, 191-195.

[114] Shan, C.; H. Yang. (2009). Direct electrochemistry of glucose oxidase and biosensing for glucose based on graphene. *Analytical Chemistry,* 81, 2378-2382.

[115] Zhang, Q.; S. Wu. (2011). Fabrication of polymeric ionic liquid/graphene nanocomposite for glucose oxidase immobilization and direct electrochemistry. *Biosensors and Bioelectronics,* 26, 2632-2637.

[116] Chen, X.; H. Ye. (2010). Electrochemiluminescence biosensor for glucose based on graphene/nafion/GOD film modified glassy carbon electrode. *Electroanalysis*, 22, 2347-2352.

[117] Liu, J.; N. Kong. (2013). Graphene bridged enzyme electrodes for glucose biosensing application. *Analyst*, 138, 2567-2575.

[118] Liu, S.; J. Tian. (2011). Self-assembled graphene platelet–glucose oxidase nanostructures for glucose biosensing. *Biosensors and Bioelectronics*, 26, 4491-4496.

[119] Mohanty, N.; V. Berry. (2008). Graphene-based single-bacterium resolution biodevice and DNA transistor: interfacing graphene derivatives with nanoscale and microscale biocomponents. *Nano Letters,* 8, 4469-4476.

[120] Bonanni, A.; M. Pumera. (2011). Graphene platform for hairpin-DNA-based impedimetric genosensing. *ACS Nano*, 5, 2356-2361.

[121] Bo, Y.; H. Yang. (2011). A novel electrochemical DNA biosensor based on graphene and polyaniline nanowires. *Electrochimica Acta,* 56, 2676-2681.

[122] Bo, Y.; W. Wang. (2011). A DNA biosensor based on graphene paste electrode modified with Prussian blue and chitosan. *Analyst,* 136, 1946-1951.

[123] Liu, F., J.Y. Choi. (2010). Graphene oxide arrays for detecting specific DNA hybridization by fluorescence resonance energy transfer. *Biosensors and Bioelectronics,* 25, 2361-2365.

[124] Myung, S.; A. Solanki. (2011). Graphene-encapsulated nanoparticle-based biosensor for the selective detection of cancer biomarkers. *Advanced Materials,* 23, 2221-2225.

[125] Tang, Z.; H. Wu. (2010). Constraint of DNA on functionalized graphene improves its biostability and specificity. *Small,* 6, 1205-1209.

[126] Wu, C.; Y. Zhou. (2011). A novel fluorescent biosensor for sequence-specific recognition of double-stranded DNA with the platform of graphene oxide. *Analyst,* 136, 2106-2110.

[127] Zhang, B.; T. Cui. (2011). An ultrasensitive and low-cost graphene sensor based on layer-by-layer nano self-assembly. *Applied Physics Letters,* 98, 073116-073116-3.

[128] Zhang, B., Q. Li. (2012). Ultra-sensitive suspended graphene nanocomposite cancer sensors with strong suppression of electrical noise. *Biosensors and Bioelectronics,* 31, 105-109.

[129] Kafi, A., G. Wu. (2008). A novel hydrogen peroxide biosensor based on the immobilization of horseradish peroxidase onto Au-modified titanium dioxide nanotube arrays. *Biosensors and Bioelectronics,* 24, 566-571.

[130] Zhou, K.; Y. Zhu. (2011). Preparation and application of mediator-free H_2O_2 biosensors of graphene-Fe_3O_4 composites. *Electroanalysis,* 23, 862-869.

[131] Zhiguo, G.; Y. Shuping. (2011). An ultrasensitive hydrogen peroxide biosensor based on electrocatalytic synergy of graphene–gold nanocomposite, CdTe–CdS core–shell quantum dots and gold nanoparticles. *Analytica Chimica Acta,* 701, 75-80.

[132] Lu, Q.; X. Dong. (2010). Direct electrochemistry-based hydrogen peroxide biosensor formed from single-layer graphene nanoplatelet–enzyme composite film. *Talanta,* 82, 1344-1348.

[133] Zeng, Q.; J. Cheng. (2010). Self-assembled graphene-enzyme hierarchical nanostructures for electrochemical biosensing. *Advanced Functional Materials,* 20, 3366-3372.

[134] Wang, Y.; Y. Shao. (2010). Nitrogen-doped graphene and its application in electrochemical biosensing. *ACS Nano*, 4, 1790-1798.

[135] Wu, L.; H. Chu. (2010). Highly sensitive graphene biosensors based on surface plasmon resonance. *Optics express*, 18, 14395-14400.

[136] Xiao, F.; Y. Li. (2012). Growth of metal–metal oxide nanostructures on freestanding graphene paper for flexible biosensors. *Advanced Functional Materials,* 22, 2487-2494.

[137] Yang, K.; J. Wan. (2010). In vivo pharmacokinetics, long-term biodistribution, and toxicology of PEGylated graphene in mice. *ACS Nano,* 5, 516-522.

[138] Dey, R.S.; C.R. Raj. (2010). Development of an amperometric cholesterol biosensor based on graphene− Pt nanoparticle hybrid material. *The Journal of Physical Chemistry C*, 114, 21427-21433.

[139] Li, H.; Q. Wei. (2011). Electrochemical immunosensors for cancer biomarker with signal amplification based on ferrocene functionalized iron oxide nanoparticles. *Biosensors and Bioelectronics,* 26, 3590-3595.

[140] Feng, L.; Y. Chen. (2011). A graphene functionalized electrochemical aptasensor for selective label-free detection of cancer cells. *Biomaterials*, 32, 2930-2937.

[141] Yang, M.; Y. Kostov. (2008). Carbon nanotubes with enhanced chemiluminescence immunoassay for CCD-based dctcction of staphylococcal enterotoxin B in food. *Analytical Chemistry,* 80, 8532-8537.

[142] Huang, Y.; X. Dong. (2011). Graphene-based biosensors for detection of bacteria and their metabolic activities. *Journal of Materials Chemistry,* 21, 12358-12362.

[143] Ohno, Y., K. Maehashi. (2010). Label-free biosensors based on aptamer-modified graphene field-effect transistors. *Journal of the American Chemical Society,* 132, 18012-18013.

[144] Lee, J.S.; H.-A. Joung. (2012). Graphene-based chemiluminescence resonance energy transfer for homogeneous immunoassay. *ACS Nano,* 6, 2978-2983.

[145] Mao, K.; D. Wu. (2012). Label-free electrochemical immunosensor based on graphene/methylene blue nanocomposite. *Analytical Biochemistry,* 422, 22-27.

[146] Wan, Y.; Y. Wang. (2010). Graphene oxide sheet-mediated silver enhancement for application to electrochemical biosensors. *Analytical Chemistry,* 83, 648-653.

[147] Liu, F.; K.S. Choi. (2011). Graphene-based electrochemical biosensor for pathogenic virus detection. *BioChip Journal,* 5, 123-128.

[148] Rodríguez, M.C.; G.A. Rivas. (2009). Label-free electrochemical aptasensor for the detection of lysozyme. *Talanta*, 78, 212-216.

[149] Nishikawa, T.; D. Edelstein. (2000). Normalizing mitochondrial superoxide production blocks three pathways of hyperglycaemic damage. *Nature*, 404, 787-790.

[150] Shan, C.; H. Yang. (2010). Electrochemical determination of NADH and ethanol based on ionic liquid-functionalized graphene. *Biosensors and Bioelectronics,* 25, 1504-1508.

[151] Tang, L.; Y. Wang. (2009). Preparation, structure, and electrochemical properties of reduced graphene sheet films. *Advanced Functional Materials,* 19, 2782-2789.

[152] Wang, K.; Q. Liu. (2011). A highly sensitive and rapid organophosphate biosensor based on enhancement of CdS–decorated graphene nanocomposite. *Analytica Chimica Acta,* 695, 84-88.

[153] Wang, K.; H.-N. Li. (2011). TiO_2-decorated graphene nanohybrids for fabricating an amperometric acetylcholinesterase biosensor. *Analyst*, 136, 3349-3354.

[154] Choi, B.G.; H. Park. (2010). Solution chemistry of self-assembled graphene nanohybrids for high-performance flexible biosensors. *ACS Nano,* 4, 2910-2918.

[155] Yue, R., Q. Lu. (2011). A novel nitrite biosensor based on single-layer graphene nanoplatelet–protein composite film. *Biosensors and Bioelectronics,* 26, 4436-4441.

[156] Wang, L.; X. Zhang. (2010). A novel nitromethane biosensor based on biocompatible conductive redox graphene-chitosan/ hemoglobin/ graphene/ room temperature ionic liquid matrix. *Biosensors and Bioelectronics,* 26, 991-995.

[157] Wu, J.-F., M.-Q. Xu. (2010). Graphene-based modified electrode for the direct electron transfer of cytochrome c and biosensing. *Electrochemistry Communications*, 12, 175-177.

[158] Liu, M.; H. Zhao. (2011). A "turn-on" fluorescent copper biosensor based on DNA cleavage-dependent graphene-quenched DNAzyme. *Biosensors and Bioelectronics,* 26, 4111-4116.

[159] Zhao, X.-H.; R.-M. Kong. (2011). Graphene–DNAzyme based biosensor for amplified fluorescence "Turn-On" detection of Pb^{2+} with a high selectivity. *Analytical Chemistry,* 83, 5062-5066.

[160] Hou, S.; M.L. Kasner. (2010). Highly sensitive and selective dopamine biosensor fabricated with silanized graphene. *The Journal of Physical Chemistry C*, 114, 14915-14921.
[161] Ge, S.; M. Yan. (2012). Electrochemical biosensor based on graphene oxide–Au nanoclusters composites for L-cysteine analysis. *Biosensors and Bioelectronics*, 31, 49-54.
[162] Liu, K.; J. Zhang. (2010). Direct electrochemistry and electrocatalysis of hemoglobin based on poly (diallyldimethylammonium chloride) functionalized graphene sheets/room temperature ionic liquid composite film. *Electrochemistry Communications,* 12, 402-405.
[163] Zhang, X.; S. Li. (2011). A new photoelectrochemical aptasensor for the detection of thrombin based on functionalized graphene and CdSe nanoparticles multilayers. *Chemical Communications*, 47, 4929-4931.
[164] Liu, F.; Y. Piao. (2012). Fabrication of free-standing graphene composite films as electrochemical biosensors. *Carbon*, 50, 123-133.
[165] Qu, Y.; M. Ma. (2013). Sensitive amperometric biosensor for phenolic compounds based on graphene–silk peptide/tyrosinase composite nanointerface. *Biosensors and Bioelectronics,* 44, 85-88.
[166] Mao, S.; G. Lu. (2010). Specific protein detection using thermally reduced graphene oxide sheet decorated with gold nanoparticle-antibody conjugates. *Advanced Materials,* 22, 3521-3526.
[167] Sheng, L.; J. Ren. (2011). PVP-coated graphene oxide for selective determination of ochratoxin A via quenching fluorescence of free aptamer. *Biosensors and Bioelectronics,* 26, 3494-3499.

In: Graphene
Editor: Bruce T. Edwards
ISBN: 978-1-62948-795-3

Chapter 2

ADSORPTION OF HEXAVALENT CHROMIUM FROM AQUEOUS SOLUTIONS BY GRAPHENE MODIFIED WITH CETYLTRIMETHYLAMMONIUM BROMIDE

Hanjin Luo* and Yan Wu
College of Environment and Energy,
South China University of Technology, Guangzhou, Guangdong, China
The Key Laboratory of Pollution Control and Ecosystem Restoration
in Industry Clusters of Ministry of Education, Guangzhou, China

ABSTRACT

In this study, cetyltrimethylammonium bromide was chosen to modify graphene which was prepared using a modified Hummers' method. The characteristics of graphene and modified graphene were characterized by X-ray diffraction, fourier transform infrared spectrum, X-ray photoelectron spectroscopy, transmission electron microscopy and scanning electron microscopy. The effect factors including pH, contact time, temperature and dosage on the adsorption properties of Cr(VI) onto graphene and modified graphene were investigated. Batch experiments were conducted to evaluate the adsorbance of Cr(VI) from aqueous

* Corresponding author. Address: College of Environment and Energy, South China University of Technology, Guangzhou 510006, Guangdong, China. Tel.: +86 20 87110517; fax: +86 20 3938 0508. E-mail address: luohj@scut.edu.cn.

solution using graphene and modified graphene as the adsorbent under different conditions. The results revealed that the optimal pH for the adsorption was about 2, and the best suitable temperature was at 293 K. The adsorption processes were rapid within the first 5 min and reached equilibrium in about 40 min. The adsorption kinetics fitted well with pseudo-second-order model. The adsorption capacity of Cr(VI) on modified graphene inferred from the Langmuir model was 21.57 mg/g at 293 K. The thermodynamic parameters indicated that the adsorption of Cr(VI) onto modified graphene was an exothermic and spontaneous process.

Keywords: Graphene, cetyltrimethylammonium bromide, adsorption, hexavalent chromium, characteration

1. Introduction

Hexavalent chromium in discharged wastewater typically comes from industrial pollution sources, including tanning factories, steel works, wood preservation and artificial fertilizers. It is widely recognized that Cr(VI) has harmful effects on the environment and human health [1]. It is acknowledged by worldwide people that Cr(VI) is carcinogenic and can be easily absorbed into the human body through digestive system, respiratory tract and skin contact. Cr(VI) can be easily absorbed into the human body through digestive system, respiratory tract and skin contact which is carcinogenic to human recognized by worldwide [2,3]. The maximum permissible concentration of the total chromium in drinking water has been identified as 100 μg/L by the U.S. Environmental Protection Agency (EPA) [4]. As Cr(VI) contamination is a widespread environmental problem, the technologies to remove Cr(VI) have been developed, including cyanide treatment [5], electro-chemical precipitation [6], reverse osmosis (RO) [7,8], ion exchange [9,10], and adsorption [11-18]. Currently, adsorption of Cr(VI) onto different adsorbents is considered as one of the most promising treatment approaches for Cr(VI) polluted wastewater [19], because this method has a large number of prominent advantages such as low-cost, flexibility and simplicity of design, ease of operation, insensitivity to toxic pollutants and avoidance of the formation of secondary pollutants [20].

Graphene (GN) is an eminent new member of carbon materials with a novel one-atom-thick two-dimensional structure [21,22]. At present, GN has received intensive attentions and is widely applied at many technological fields such as field-effect transistors [23-25], solar cell [26], sensor [27], supercapacitors [28] and transparent electrodes [29] due to its exceptional properties such as the fascinating mechanical, electrical, thermal and optical properties. Usually, the synthesis of most GN-based multifunctional hybrid materials stems from chemically oxidized graphene oxide (GO). GO has copious oxygen-containing groups, such as epoxides, alcohols, lactols, ketones, and carboxyl groups, which can significantly affect the van der Waals interactions between the graphene layers [30]. Moreover, GO can be readily made from low-cost natural graphite in large scale. Therefore, hybrid multifunctional materials based on GO are much more applicable than those based on the other more expensive nanomaterials [31]. Recently, experimental results reveal that GN is a prominent adsorbent to remove heavy metals [32,33], dyes [34,35] and fluoride ions [36]from aqueous solutions.

Chemical functionalization of GN can be introduced through covalent modification or noncovalent decoration [37]. In this study, CTAB, a cationic surfactant, has been chosen to modify GN *via* noncovalent interaction. The negatively charged carboxyl group of GO can interact with the positive charged ammonium ion of CTAB through ionic interactions, and then the CTAB-GO is reduced to form CTAB-GN [38]. Previous research indicated that modification of some adsorbents with CTAB could greatly enhanced the sorption of pollutants [39]. For example, Jin et al. [40] found that the modification of Fe_3O_4 with CTAB greatly increased As(V) adsorption capacity. Moreover, chemical functionalization of GN with CTAB also prevents the agglomeration of single layer GN and helps to maintain the inherent properties of GN, such as large surface area, good dispersibility, long-term durability, and so on [38]. Therefore, the presence of CTAB in GN materials is expected to enhance the adsorptive performance for the removal of Cr(VI). However, the adsorption of Cr(VI) on GN modified with CTAB has never been explored in detail and thus requires investigation. The objectives of this study are: (1) to prepare modified graphene(CTAB-GN) and apply them as sorbents to remove Cr(VI) ions from aqueous solutions; (2) to analyze the structures of CTAB-GN and explore the adsorption mechanism; (3) to investigate the effects of pH, contact time, temperature and dosage on Cr(VI) adsorption.

2. Materials and Methods

2.1. Materials

All chemicals and reagents were of analytical grade in the experiments. All solutions were prepared using deionized (DI) water. The initial Cr(VI) solution was prepared by dissolving $K_2Cr_2O_7$ in deionized water.

2.2. Methods

2.2.1. Preparation of GN and CTAB-GN

GO was synthesized from natural graphite (Qingdao Nanshu Ruiying Graphite Co. Ltd., China) by a modified Hummers method [41]. Graphite (5.0 g) and $NaNO_3$ (2.5 g) were mixed with 120mL of H_2SO_4 (98%) in a 500mL flask. The mixture was stirred for 30 min in ice bath. While maintaining vigorous stirring, potassium permanganate (15.0 g) was added to the suspension. The rate of addition was carefully controlled to keep the reaction temperature lower than 288 K. After 90 min, the ice bath was removed and the mixture was stirred at 308 K for 30 min. As the reaction progressed, the mixture gradually became pasty and the color turned into light brownish. At the end, 230 mL of H_2O was slowly added to the paste with vigorous agitation. The reaction temperature was rapidly increased to 371 K, and the color changed into yellow. The diluted suspension was stirred at 371 K for 30min. Then, 30 mL of 30% H_2O_2 was added to the mixture. For purification, the mixture was washed by rinsing and centrifugation with 5% HCl then DI water several times. After filtration and drying under vacuum, GO was obtained as grey powder. GN was synthesized by sodium borohydride reduction of GO. The dispersion of graphite oxide was ultrasonicated until no visible particles existing. Then, sodium borohydride was added to the dispersion and heated at 353 K for 2 h. The final product was filtered and then washed thoroughly with deionized water, and dried at 323 K. The CTAB-GN was prepared by dispersing 0.5 g GO in 250 mL CTAB solution (0.04M). This solution was kept stirring for 2 h at 353 K. Finally, the product was filtered and then washed thoroughly with deionized water, and dried at 323 K.

2.2.2. Characterization of GN and CTAB-GN

Transmission electron microscope (TEM) images were obtained using a JEOL (Japan) JEM-100CX-II TEM operated at accelerating voltage 100 kV. The freshly prepared CTAB-GN and GN samples were dispersed by ultrasonic before test. Scanning electron microscope (SEM) images were obtained using a JEOL (Japan) S-3700N SEM in different magnification with an accelerating voltage of 15 kV.

The X-ray diffraction (XRD) patterns were recorded using a Bruker D8-advance X-ray diffractometer at 40 kV and 40 mA. A Cu Kα radiation (λ =1.5418 Å) source was used in the X-ray tests. All patterns were collected at a scanning rate of 0.020 s/step and 17.7 s/step, 2θ ranging from 5° to 70°. The sample was grounded and mounted on a flat sample plate. In order to mitigate the effects of preferred orientation that might distort the observed diffraction patterns, the powdered sample sieved manually to give particle size ranges of below 48μm [42].

Fourier transform infrared spectrum (FTIR) patterns were determined using a Bruker Tensor 27 FTIR spectrophotometer. Typically, 100 scans were collected at a resolution of 4 cm^{-1} in the range of 4000–400 cm^{-1}. The X-ray photoelectron spectroscopy (XPS) measurements were conducted with an Axis UltraDLD spectrometer (Kratos Analytical Ltd., England) with amonochromatized Al KαX-ray source (hν=1486.6eV). Thermogravimetric analysis (TGA) curves were measured by using a SDT Q600 thermogravimetric analyzer from room temperature to 800°C with heating rate of 10°C/min and an air flow rate of 100 mL/min under nitrogen flow.

2.2.3. Cr(VI) Adsorption Experiments

The batch experiments of Cr(VI) sorption on GN and CTAB-GN were conducted in triplicate in a mechanical shaker at an agitation speed of 150 rpm. Kinetics of adsorption at room temperature were studied by adding 0.2 g adsorbents in a series of flasks. Then, 25 ml of Cr(VI) at initial concentration of 50 ppm and 100 ppm were added into each flask and the mixture was stirred for a predetermined time intervals (1, 3, 5, 10, 20, 30, 40, 60, 90 and120 min).

The effect of adsorbent dose on the adsorption of Cr(VI) was studied by shaking 50 ml of 50 ppm solutions of Cr(VI) with varying adsorbent dosages (0.1–1.0 g) for 60 min. The effect of pH on adsorption of Cr(VI) was studied in a pH range of 1.0–12.0 by agitating adsorbent (0.4 g) with Cr(VI) (50 mL, 50 ppm) for 60 min at 150 rpm. The pH was adjusted by adding aqueous solutions of 0.1M HCl or 0.1M NaOH. Isothermal studies at different temperatures (293 K, 313 K, and 333 K) were carried out by adding 0.2 g adsorbent into 50 ml of Cr(VI) solution of varying concentrations from 20 to 100 ppm.

In all above cases, after shaking, the samples were centrifuged for 5 min, then filtered through 0.45 μm filter membrane. The filtrate was applied as the analysis of remaining Cr(VI) in the solution.

2.2.4. Determination of Cr (VI) Concentration

The concentration of chromium was measured using a UV-vis Spectrophotometer (UV-1750, SHIMADZU, Japan), which analyzed by the purple complex of Cr(VI) with 1,5-diphenylcarbazide at 540 nm [43]. The specific adsorbed amount of Cr(VI) was calculated according to the following equation:

$$q_e = \frac{C_0 - C_e}{V} \times m \tag{1}$$

Where C_0 and C_e are the initial and equilibrium concentrations of Cr(VI) in solution (ppm), V is the volume of solution (L), m is the mass of adsorbent (g).

3. Results and Discussions

3.1. Characterizations of GN and CTAB-GN

3.1.1. Transmission Electron Microscopy

The morphological structure of GN and CTAB-GN is characterized by TEM and shown in Figure 1. Figure 1(a) shows the GN layer displays stacking disorder. It is mainly because in the oxidation process, the introduction of the hybrid carbon atoms leads to the disturbances of planar sp^2 carbon layer. In addition, in order to ensure thermal stability, the thin-layer GN has prompted its spontaneous stacking and curling features. From Figure 1(b), we can see that CTAB-GN exhibits a curved thin flaky appearance and shows many wrinkled and folded regions. The TEM images of CTAB-GN also shows that there are no regularly arranged layer-like structures, even at the wrinkled edge part. The disordered or irregular stacked layerd structure of CTAB-GN indicates that GN sheets in single or few layers structure are stabilized by surfactants during the reduction treatment [38].

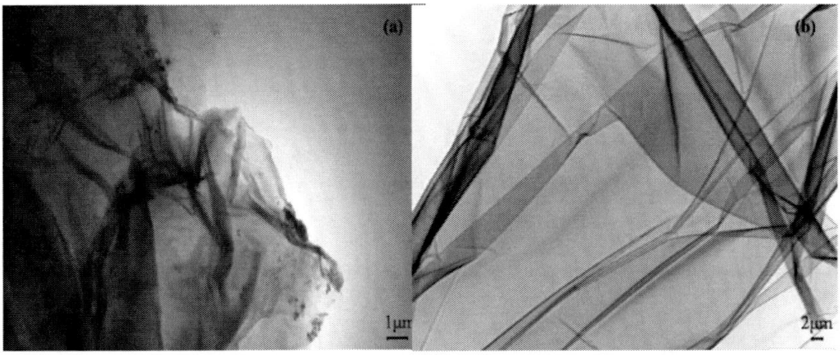

Figure 1. The TEM images of GN (a) and CTAB-GN (b).

3.1.2. Scanning Electron Microscopy

The morphology of GN and CTAB-GN is observed by SEM. From Figure 2(a), we can clearly observe in thin-layer GN that the spontaneous stacking curls form petal-shaped aggregates. This flocculent aggregate in the absence of external forces will remain stable and it is difficult to separate from each other. Figure 2(b) shows that CTAB-GN has flake structure and the size is also larger than the graphene. This is mainly due to the CTAB makes the graphene layer peeling off more obvious, graphene layers become thinner, thus increasing the adsorption surface area.

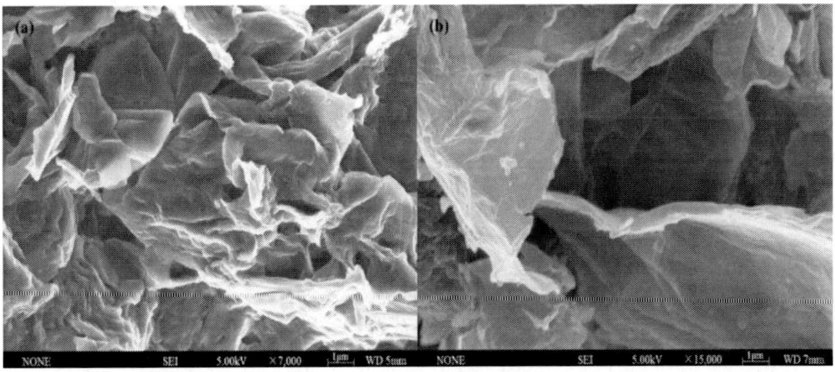

Figure 2. The SEM images of GN (a) and CTAB-GN (b).

3.1.3. X Ray Diffraction

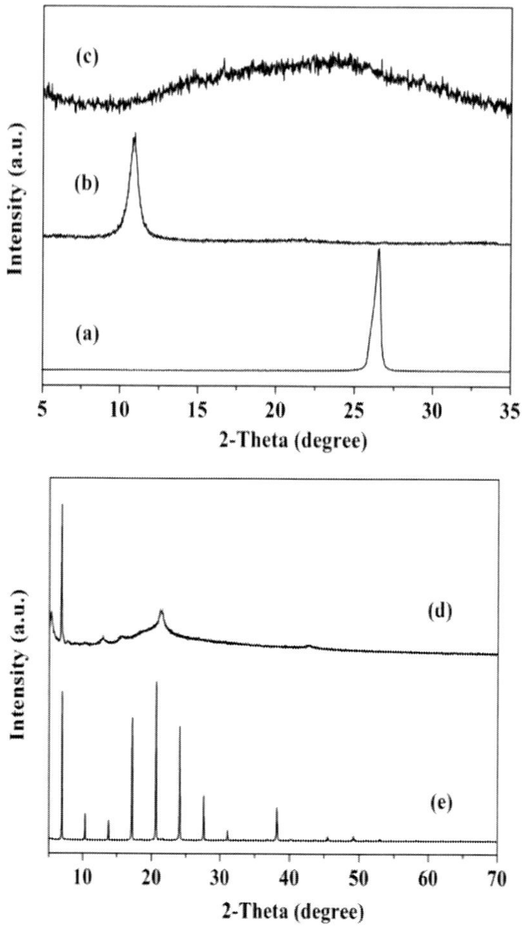

Figure 3. The XRD patterns of graphite(a), GO(b), GN(c), CTAB-GN(d) and CTAB(e).

Figure 3 shows the XRD patterns of the graphite, GO, GN, CTAB-GN and CTAB. Graphite has a sharp diffraction peak around $2\theta = 26.6^{\circ}$, indicates an interlayer spacing of 0.334 nm, which corresponds to the normal graphite spacing of graphite plane. After oxidation, the characteristic graphite peak disappears and is replaced by a well-defined peak at $2\theta = 10.9^{\circ}$ corresponding to an interlayer spacing of 0.808 nm, which indicates a typical layer-like structure [31]. The d-spacing increases from 0.334 to 0.808nm after the graphite is modified to GO sheets, which ascribed to the presence of abundant oxygen-containing functional groups on the surfaces of the GO sheets [44]. In contrast, the XRD pattern of GN shows a wide peak at $2\theta = 24.5°$ with 0.362 nm d-spacing, which is much lower than that of GO, confirming the recovery of ordered graphitic crystal structure after reduction by sodium borohydride. Besides the peak at 24.5° corresponding to the GN, an additional peak at 2θ values of 7.1° can be clearly seen from the XRD patterns of the CTAB-GN, indicating the coexistence of CTAB and GN in the composite, which is also supported by the FT-IR measurements (to be discussed in the following part.

3.1.4. Fourier Transform Infrared Spectrum

Figure 4 shows the FTIR spectra between 4000 and 400 cm^{-1} for GN and CTAB-GN. There are characteristic peaks of CTAB emerging on infrared spectra of CTAB-GN, though the stretching intensity is much weaker. The band at 1558 cm^{-1} indicates the existence of the N-H bending vibration absorption peak which explains the formation of the amide bond, namely amino groups in the CTAB and carboxy groups in the GN have generated amide bond. These results indicate that CTAB modifies the surface spacing of GN successfully.

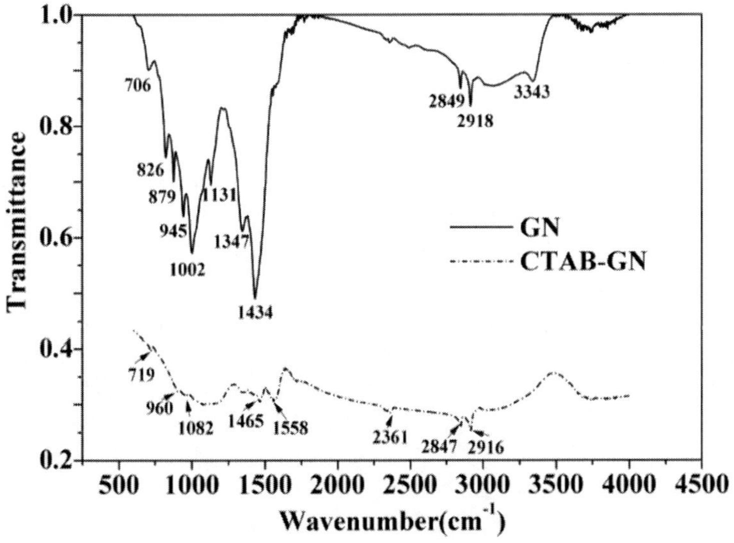

Figure 4. The FTIR spectrums of GN and CTAB-GN.

3.1.5. X-Ray Photoelectron Spectrum

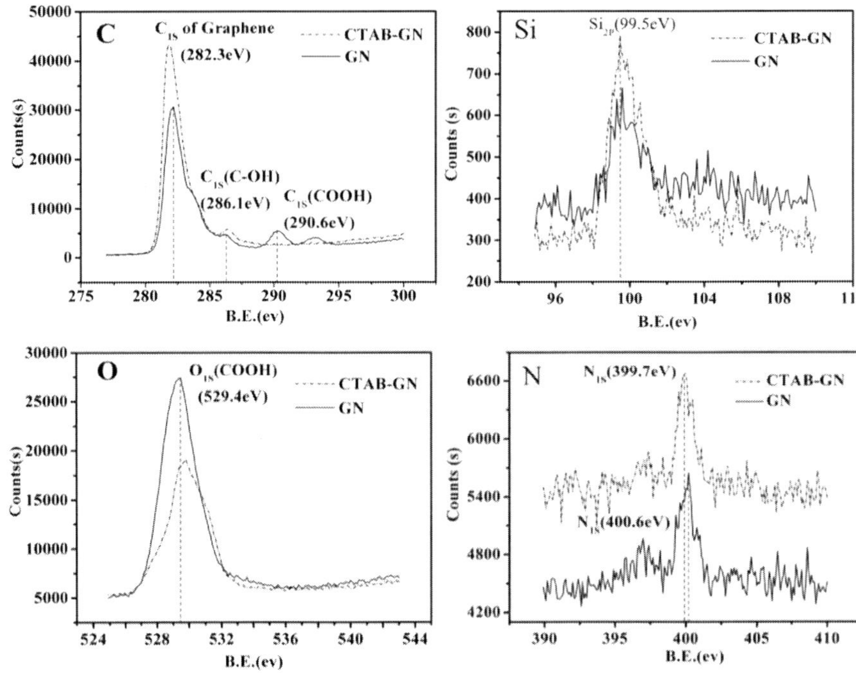

Figure 5. The XPS spectrums of Si, N, O and C at GN and CTAB-GN.

The element characterization of powders of GN and CTAB–GN was analyzed by XPS separately. The Si_{2p} peaks(Figure 5) are associated with the binding energy of 99.9 eV represents the bond of silicon with -OH from the GN, resulting from the partial hydrolysis of the silane molecules during the silylation reaction [45]. The N_{1s} XPS spectrum of GN contains one peak at 400.6 eV, which is assigned to $N–C^{3}_{sp}$ or $N–C^{2}_{sp}$. Nitrogen is introduced to the graphene surface during the synthesis and reduction process [46]. Meanwhile, two peaks of N_{1s} are observed in the CTAB-GN sample: one is the same as that for GN, and the other at 399.7eV is presumably due to the tertiary amines of the CTAB [47]. The O_{1s} spectra (Figure 5) can be fitted with the peak at binding energy of 529.4 eV for both GN and CTAB–GN. The O_{1s} peak at 529.4 eV is likely contributed from the C=O or O=C–OH groups [48]. In comparison to the O_{1s} spectrum of GN, the CTAB–GN sample clearly exhibits a peak with increased intensity corresponding to the carboxyl groups. The C_{1s} spectra (Figure 5) of GN and CTAB–GN both contains three components of the carbon in C–C at 284.6 eV, the carbon in C–OH at 286.1 eV and the carbonyl carbon (C=O) at 290.6 eV [49]. The peak intensity at 290.6 eV of CTAB–GN is much stronger than that of GN, indicating a considerable proportion of the COOH groups (from CTAB) are linked to the GN surface [50].

3.1.6. Thermogravimetric Analysis

The compositions of GN and CTAB-GN are studied by TGA. As shown in Figure 6, GN exhibit a smooth weight loss from 100°C to 800°C, with 27% total loss at 800°C. CTAB-GN shows significant weight loss at around 200°C, which may be ascribed to the decomposition of CTAB in the interlayer of GN. The TGA curve of CTAB-GN has 74% weight loss at 800°C. Thus, the contents of surfactant in CTAB-GN can be roughly estimated to be 47%.

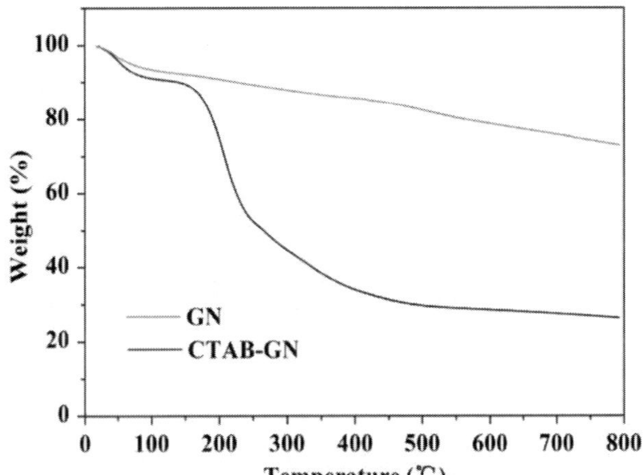

Figure 6. The TGA curves of GN and CTAB-GN.

3.2. Effect of Contact Time

The effect of contact time on the adsorption of Cr(VI) onto GN and CTAB-GN is carried out at 298 K and shown in Figure 7. It is obvious that the adsorption of Cr (VI) onto GN and CTAB-GN have the similar trend, but the adsorption capacity of Cr(VI) onto CTAB-GN is much greater than the adsorption capacity of Cr(VI) onto GN. It can be seen that the adsorption of Cr(VI) onto GN and CTAB-GN increases quickly within 5 min. This result might be due to the availability of large number of vacant sites that saturated with time. Meanwhile, the special layered structure of these materials is also helpful to the removal of Cr(VI). With further increasing time, the change of adsorptive capacity for GN and CTAB-GN tend to slow and gradually reach the equilibrium state in 40 min for both concentration, which might ascribe to the diminishing availability of the remaining active sites and the decrease in the driving force [51].According to this result, we select 1 h as an optimum equilibrium time for the dosage, pH effects and adsorption isotherm experiments.

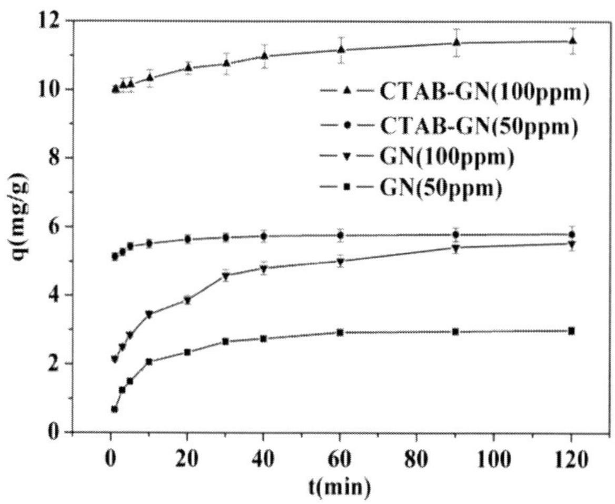

Figure 7. Effect of contact time on the adsorption of Cr(VI) at different initial concentrations.

3.3. Effect of Adsorbent Dose

The relation between the dosage of the GN and CTAB-GN and the removal efficiency was carried out, and the result was shown in Figure 8. It is obvious that the Cr(VI) removal percentage of GN and CTAB-GN increases with the increase of adsorbent dose until the dosage is up to 0.6 g and 0.4 g, respectively. The removal ratio of the Cr(VI) correspondingly increases from 55.9% to 73.5% for GN, and 78.3% to 98.2% for CTAB-CN. The reason might be due to that increasing adsorbent dosage does assist in enhancing the surface area and the number of active sites for adsorption [52]. Above 0.6g and 0.4g of GN and CTAB-GN dosage, respectively, the adsorption equilibrium of the Cr(VI) is reached and the removal ratio of Cr(VI) hold almost constant. The dosage of CTAB-GN is set at 0.4 g in the following experiments according to the above results.

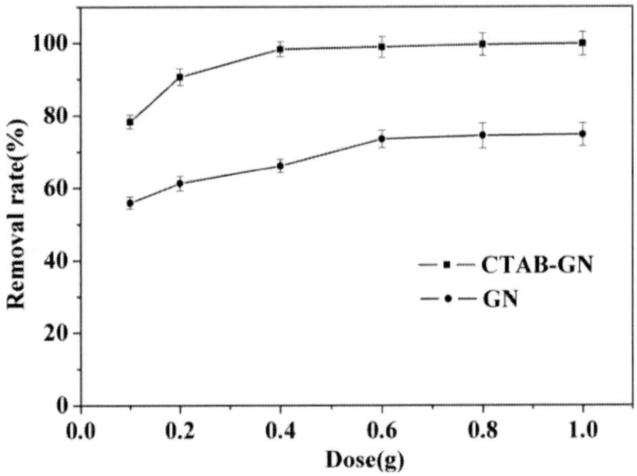

Figure 8. Effect of amount of adsorbent on the adsorption of Cr(VI).

3.4. Effect of pH

The pH of aqueous solution is one of the most important factors to study the adsorption property of an adsorbent [53]. Figure 9 shows the influence of the pH on the adsorption of Cr(VI) onto GN and CTAB-GN. The pH of solution plays a key role in the adsorption of Cr(VI) onto GN and CTAB-GN. We find that the adsorption capacity of Cr(VI) is highly dependent on the pH of solution. The maximum removal rate of Cr(VI) is achieved at the pH of 2. Therefore, CTAB-GN exhibits much higher adsorption capacity under strong acidic condition rather than in neutral and alkaline conditions. Cr(VI) species have many different ionic forms in solution. Chromate (CrO_4^{2-}), dichromate ($Cr_2O_7^{2-}$) and hydrogen chromate ($HCrO_4^-$) are the main Cr(VI) ion forms in solution and these ion forms are related to the solution pH and total chromate concentration [54-57]. The major species of Cr(VI) are CrO_4^{2-} and $HCrO_4^-$ which infers from the predominance diagram of the chromium species based on thermodynamic database [58-59] using both pH and total Cr(VI) as variables. For pH lower than 6.8, $HCrO_4^-$ is the predominant species and at pH above 6.8, only CrO_4^{2-} is stable. Results show that GN is probably more selective for the adsorption of $HCrO_4^-$ than CrO_4^{2-} due to the surface chemical reaction between GN and Cr(VI) depends highly on pH in solution. The surface of metal oxides is generally covered with hydroxyl groups that vary in form at different pH levels. The uptake of Cr(VI) ions decreased with the increase in pH, which is due to the higher concentration of OH^- ions present in the mixture compete with Cr(VI) species [60].

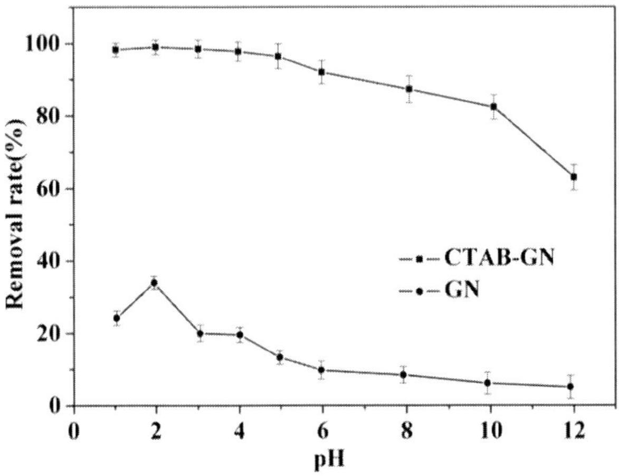

Figure 9. Effect of pH on the adsorption of Cr(VI) to GN and CTAB-GN.

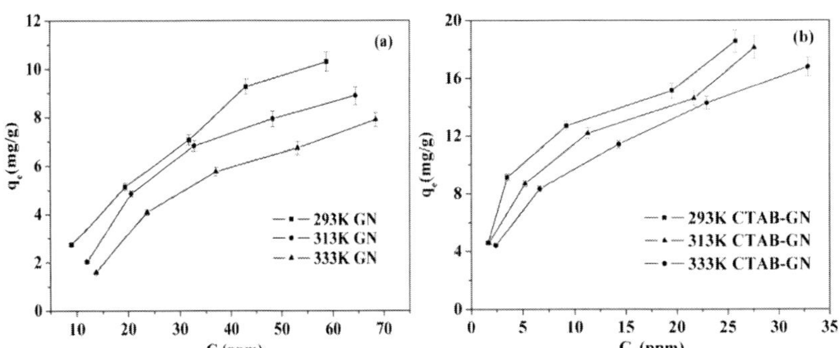

Figure 10. Effect of temperature on the adsorption of Cr(VI) to GN(a) and CTAB-GN(b).

3.5. Effect of Temperature

The temperature is another important factor to influence the process of adsorption. The effect of temperature on adsorption of Cr (VI) onto GN and CTAB-GN has been investigated at 293, 313 and 333 K, and the results are presented in Figure 10. It can be seen that the adsorption capacity decreases as the temperature increases. When the temperature of solution increases, the surfactant is desorbed from the CTAB-GN surface [61]. This suggests that the lower temperatures is good for adsorption. The experimental results demonstrate that the process of adsorption of Cr (VI) onto GN and CTAB-GN is exothermic.

3.6. Adsorption Isotherm Study

Adsorption isotherms can explain the reaction between adsorbate and adsorbent, providing the most important parameter for designing a desired adsorption system. The Freundlich and Langmuir isotherms models are conventional models to fit the experimental data. In this study, both models are applied to describe the experimental data obtained at three temperatures (293, 313 and 333 K). The Langmuir isotherm is often applicable to a homogeneous adsorption surface with all the adsorption sites having equal adsorbate affinity. The equation of the Langmuir isotherm is as follows [62]:

$$\frac{C_e}{q_e} = \frac{C_e}{q_{\max}} + \frac{1}{q_{\max}} \cdot k_L \tag{2}$$

where q_e is the amount adsorbed at equilibrium (mg/g), C_e is the equilibrium concentration of the solution (ppm), q_{max} is the maximum adsorption capacity (mg/g), k_L is a Langmuir constant related to the affinity of the binding sites and energy of adsorption (L/g). A straight line is obtained when C_e/q_e is plotted against C_e which is shown in Figure 11(a), q_{max} and k_L could be evaluated from the slope and intercept (Table 1). The maximum monolayer adsorption capacity (q_m) of the CTAB-GN composite is calculated to be 21.59 mg/ g. The high value of coefficient of determination indicates a good agreement between the experimental values and isotherm parameters and also confirms the monolayer adsorption of Cr(VI) onto the composite surface. Moreover, the essential feature of the Langmuir isotherm can be expressed in terms of a dimensionless constant separation factor (R_L) given by the following equation [63]:

$$R_L = \frac{1}{1 + k_L C_0} \tag{3}$$

Where k_L is the Langmuir constant (L/g) and C_0 is the highest initial concentration (ppm) in the liquid phase. This value of parameter indicates the shape of isotherm is unfavorable ($R_L>1$), favorable ($0<R_L<1$), linear ($R_L=1$), or irreversible ($R_L=0$). For Cr(VI) adsorption on the CTAB-GN composite, R_L values obtained are in the range of 0.056–0.096, thereby confirming that the adsorption is a favorable process.

However, the Freundlich isotherm model assumes heterogeneity of adsorption surfaces. The equation is commonly expressed as follows [64]:

$$\ln q_e = \ln k_F + \frac{1}{n} \cdot \ln C_e \tag{4}$$

where k_F (L/g) is a Freundlich constant, which indicates the relative adsorption capacity of the adsorbent, *1/n* is the heterogeneity factor representing the deviation from linearity of adsorption and is also known as Freundlich coefficient. A straight line is obtained when lnq_e is plotted against lnC_e which is shown in Figure 11(b), n and k_F could be evaluated from the slope and intercept (Table 1). It can be found that the value of k_F and *n* decrease as the temperature increases, indicating that adsorption is favorable at lower temperature [65].

Table 1. Isotherm constants and regression data for various adsorption isotherms for adsorption of Cr (VI) on CTAB-GN

Temperature(K)	Langmuir				Freundich		
	q_{max}(mg/g)	k_L(L/g)	R^2	R_L	1/n	k_F (L/g)	R^2
293	21.57	0.167	0.968	0.056	0.458	4.249	0.917
313	21.59	0.129	0.958	0.072	0.474	3.689	0.980
333	21.50	0.094	0.981	0.096	0.503	2.977	0.989

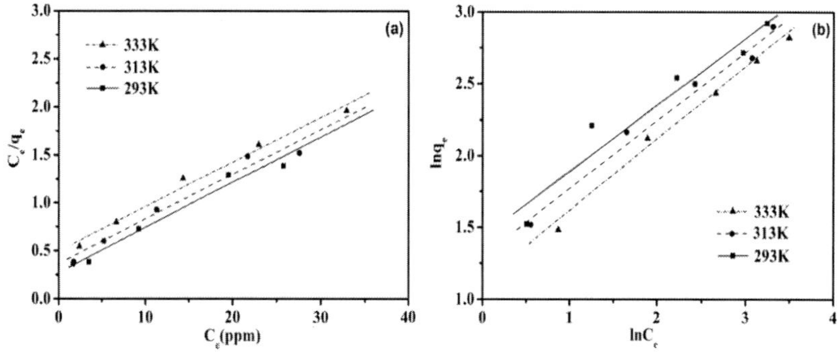

Figure 11. Langmuir isotherm model(a) and Freundlich isotherm model(b) for the Cr(VI) adsorption onto CTAB-GN.

3.7. Adsorption Kinetics

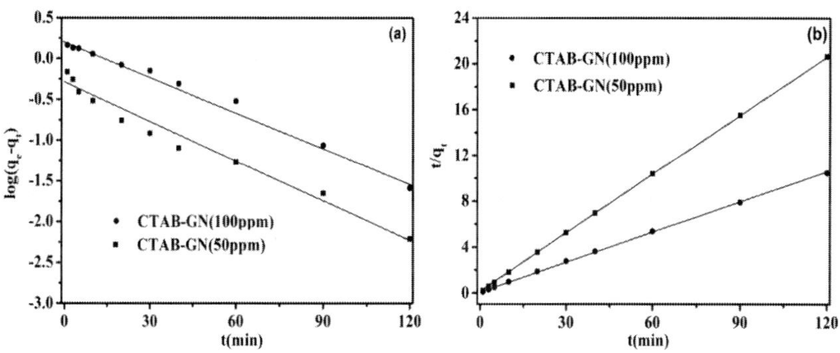

Figure 12. First-order plot(a) and Second-order plot (b)for the adsorption of Cr(VI) using CTAB-GN as an adsorbent.

In order to investigate the kinetics of adsorption of Cr(VI) onto CTAB-GN, the pseudo-first-order and pseudo-second-order kinetic models are employed in this study. The linearized-integral form of the pseudo-first-order model is expressed as [66] :

$$\log(q_e - q_t) = \log q_e - \frac{k_1 t}{2.303} \tag{5}$$

Where q_e and q_t are the amounts of Cr(VI) adsorbed (mg/g) on CTAB-GN at equilibrium time and at any instant of time t, respectively, and k_1 (L/min) is the rate constant of the pseudo first-order adsorption. As we know, k_1 is determined by plotting log (q_e–q_t) versus t (Figure 12(a)). The parameter values of the kinetic models at two initial concentrations are presented in table 2. The results tell us that the experimental q_e does not agree with calculated q_e and the coefficients of determination R^2 is low. Therefore, the adsorption of Cr(VI) onto CTAB-GN does not fit with the pseudo-first order kinetic model.

The linearized-integral form of the pseudo-second-order model is expressed as [67]:

$$\frac{t}{q_t} = \frac{1}{2} k_2 q_e^2 + \frac{t}{q_e} \tag{6}$$

where k_2 is the pseudo-second-order rate constant (g/ (mg · min)), k_2 is determined by plotting t/q_t versus t (Figure 12(b)). The parameter values of the kinetic models are listed in table 2. The calculated q_e values show a good agreement with the experimental values and the values for coefficient of determination (R^2) are more than 0.999 which indicates that the second-order kinetic model could well describe the adsorption of Cr(VI) on CTAB-GN.

Table 2. Calculated kinetic parameters for pseudo first-order and second-order kinetic models for the adsorption of Cr(VI) using CTAB-GN as an adsorbent

Concentration (ppm)	Pseudo-first-order model				Pseudo-second-order model			
	$q_{e,exp}$ (mg/g)	k_1 (1/min)	$q_{e,cal}$ (mg/g)	R^2	$q_{e,exp}$ (mg/g)	k_1 (1/min)	$q_{e,cal}$ (mg/g)	R^2
50	5.814	0.037	0.048	0.970	5.814	0.080	5.821	0.999
100	11.467	0.033	1.657	0.988	11.467	0.093	11.481	0.999

What's more, intraparticle diffusion model is applied to analyze the diffusion mechanism and identify the possible rate-controlling step. Former study showed that the adsorption of heavy metal onto GN materials might follow three steps of external diffusion, intraparticle diffusion and adsorption [68]. One or more of these steps can control the adsorption kinetics altogether or individually [69]. In a well-agitated batch system, the external diffusion resistance is much weakened. Therefore, we conjecture that intraparticle diffusion and adsorption might be the rate-controlling step. This possibility is investigated by plotting of q_t versus $t^{1/2}$, according as intraparticle diffusion model (also called Weber-Morris model), which is described by the following equation [70]:

$$q_t = K_{id} t^{\frac{1}{2}} + C \tag{7}$$

where C is the intercept and K_{id} is the intraparticle diffusion rate constant (mg/g min$^{1/2}$), which can be evaluated from the slope of the linear plot of q_t versus $t^{1/2}$ (Figure 13). Based on this model, in order to say that the intraparticle diffusion is the rate controlling step, the plot of q_t versus $t^{1/2}$ should be linear and pass through the origin [71]. In this study, as can be seen from Figure 12 and Table 3, the curves reveals linear characteristics and the linear plots do not pass through the origin. It can be concluded that intraparticle diffusion is not the only rate-controlling step, and adsorption is predominant mechanism in this process.

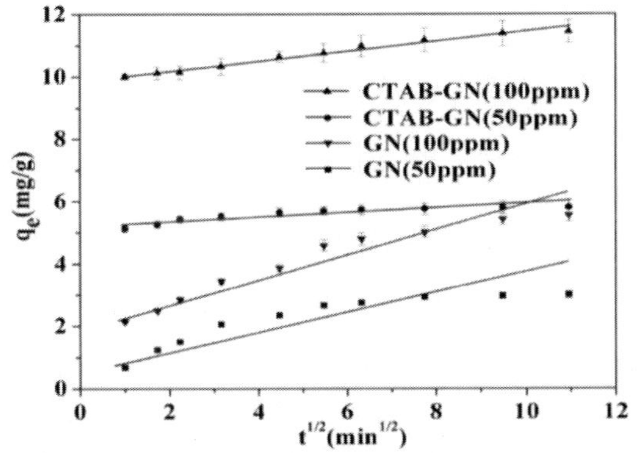

Figure 13. Intraparticle diffusion plot on Cr(VI) adsorption by GN and CTAB-GN.

Table 3. Intraparticle diffusion constants for different initial Cr(VI) concentrations.

Concentration	GN			CTAB-GN		
(ppm)	K_{id} (mg/g min$^{1/2}$)	C	R^2	K_{id} (mg/g min$^{1/2}$)	C	R^2
50	0.219	1.053	0.792	0.063	5.242	0.781
100	0.355	2.146	0.925	0.157	9.869	0.976

3.8. Thermodynamic Study

To evaluate the influence of temperature on adsorption process of Cr(VI) onto CTAB-GN, the thermodynamic parameters such as Gibbs free energy (ΔG^0), enthalpy (ΔH^0) and entropy (ΔS^0) are calculated using the following equations [72]:

$$\Delta G^0 = -RT \ln b \tag{8}$$

$$\Delta H^0 = -R\frac{T_2T_1}{T_2 - T_1}\ln(\frac{b_2}{b_1}) \tag{9}$$

$$\Delta S^0 = \frac{\Delta H^0 - \Delta G^0}{T} \tag{10}$$

where R is the universal gas constant (8.314 J/mol K), T is the absolute temperature (K), and b is Langmuir constant (the unit should be transformed into milligrams per gram for the calculation of ΔG°) at a specific temperature. The thermodynamic parameters are listed in table 3. The negative value of energy ΔG° suggests the adsorption process is feasible and spontaneous. In general, the values of energy ΔG° in between 0 and −20 kJ/mol indicate that the adsorption process is physisorption, while the values in between −80 and −400 kJ/mol correspond to chemisorptions [73,74]. The values of energy ΔG° suggest the adsorption is a physisorption and chemisorption process simultaneously. As the value of ΔH° is negative, we can infer that the adsorption reaction is exothermic. This implies that the adsorption process is energetically stable. The positive value of ΔS° reflects the increased randomness at the solid/solution interface during the adsorption process [75].

Table 4. Thermodynamic parameters for Cr(VI) adsorbed by modified graphene

T	ΔG° (kJ/mol)	ΔS° (J/(K mol))	ΔH° (kJ/mol)
293K	-29.295	59.925	
303K	-30.623	60.339	-11.737
333K	-31.703	59.958	

The value of ΔH° listed is an average one.

CONCLUSION

The introduction of CTAB groups to the GN surface can significantly increase the adsorption capacity of GN for heavy metal removal. The CTAB groups together with –OH and –COOH groups on the GN surface can make CTAB -GN an excellent adsorbent for removal of Cr (VI) in aqueous solutions. The adsorption capacity varies with different pH values of the solution. The optimal pH for the removal of Cr(VI) is about 2. The adsorption of Cr(VI) fits the Langmuir equation well. The highest adsorption capacity of CTAB-GN for Cr(VI) is found to be 21.57mg/g at 293K. The results illustrate that the adsorption of Cr(VI) onto CTAB-GN fits the pseudo-second-order model. It is found in this study that the Cr(VI) adsorption processes have reached their equilibrium state in about 40 min, which is faster than most of carbon-based adsorbents can do. The thermodynamic parameters calculated from the temperature dependent adsorption isotherms indicates that the adsorption of Cr(VI) is an exothermic and spontaneous process. This research demonstrates that CTAB-GN can be an effective adsorbent for toxic heavy metal removal.

ACKNOWLEDGMENTS

This work was financially supported by National Natural Science Foundation of China (No. 40973074) and National High Technology Research and Development Program of China (2013AA06A209).

REFERENCES

[1] Hsu, L. C., Wang, S. L., Lin, Y. C., Wang, M. K., Chiang, P. N., Liu, J. C., Kuan, W. H., Chen, C. C., Tzou, Y. M., Cr(VI) removal on fungal biomass of Neurospora crassa: The importance of dissolved organic carbons derived from the biomass to Cr(VI) reduction. *Environ. Sci.Technol.* 44(2010) 6202–6208.

[2] Alguacil F.J., Alonso M., Chromium (VI) removal through facilitated transport using CYANEX 923 as carrier and reducing stripping with hydrazine sulfate. *Environ. Sci. Technol.* 37 (2003) 1043–1047.

[3] Nami K.S., Removal of copper, chromium, and arsenic from CCA-C treated wood by EDTA extraction. *Waste Manage*. 23 (2003) 537–546.
[4] Xu Y.H., Zhao D.Y., Reductive immobilization of chromate in water and soil using stabilized iron nanoparticles. *Water Res.* 41(2007) 2101−2108.
[5] Monser, L., Adhoum, N., Modified activated carbon for the removal of copper, zinc, chromium and cyanide from wastewater. *Sep. Purif. Technol.* 26(2002)137−146.
[6] Kongsricharoern, N., Polprasert, C., Chromium removal by a bipolar electro-chemical precipitation process. *Water Sci. Technol.* 34(1996)109−116.
[7] Hafez, A., El-Mariharawy, S., Design and performance of the two stage/ two-pass RO membrane system for chromium removal from tannery wastewater. *Part 3. Desalination* 165(2004) 141−151.
[8] Modrzejewska, Z., Kaminski, W., Separation of Cr(VI) on chitosan membranes. *Ind. Eng. Chem. Res.* 38(1999)4946−4950.
[9] Rengaraj, S., Joo, C. K., Kim, Y., Yi, J., Kinetics of removal of chromium from water and electronic process wastewater by ion exchange resins: 1200H, 1500H and IRN97H. *J. Hazard. Mater.* 102(2003)257−275.
[10] Rengaraj, S., Yeon, K.H., Moon, S.H., Removal of chromium from water and wastewater by ion exchange resins. *J. Hazard. Mater.* 87(2001) 273−287.
[11] Zhang, D., Wei, S., Kaila, C., Su, X., Wu, J., Karki, A. B., Young, D. P., Guo, Z.,Carbon- stabilized iron nanoparticles for environmental remediation. *Nanoscale* 2(2010)917−919.
[12] Mohan, D., Singh, K. P., Singh, V. K., Removal of hexavalent chromium from aqueous solution using low-cost activated carbons derived from agricultural waste materials and activated carbon fabric cloth. *Ind. Eng. Chem. Res.* 44(2005)1027−1042.
[13] Sharma, Y. C., Singh, B., Agrawal, A., Weng, C. H., Removal of chromium by riverbed sand from water and wastewater effect of important parameters. *J. Hazard. Mater.* 151(2008) 789−793.
[14] Hu, J., Chen, G., M.C. Lo, I., Removal and recovery of Cr(VI) from wastewater by maghemite nanoparticles. Water Res. 39(2005)4528−4536.
[15] AydIn, Y. A., Aksoy, N. D., Adsorption of chromium on chitosan: optimization, kinetics and thermodynamics. *Chem. Eng. J.* 151(2009)188−194.

[16] Zhao, Y., Peralta-Videa, J. R., Lopez-Moreno, M. L., Ren, M., Saupe, G., Gardea-Torresdey, J. L., Kinetin increases chromium absorption, modulates its distribution, and changes the activity of catalase and ascorbate peroxidase in Mexican palo verde. *Environ. Sci. Technol.*45(2011)1082−1087.

[17] Sawalha, M. F., Gardea-Torresdey, J. L., Parsons, J. G., Saupe, G., Peralta-Videa, J. R., Determination of adsorption and speciation of chromium species by saltbush (Atriplex canescens) biomass using a combination of XAS and ICP-OES. *Microchem*. J. 81(2005) 122−132.

[18] Parsons, J. G., Hejazi, M., Tiemann, K. J., Henning, J., Gardea-Torresdey, J. L., An XAS study of the binding of copper(II), zinc(II), chromium(III) and chromium(VI) to hops biomass. *Microchem. J.* 71(2002)211−219

[19] Li J., Yue Q.Y., et al., *Adsorption of chromium (VI) from aqueous solution on bentonite modified by cationic polymers*, Huanjing Kexue 30 (2009) 1738–1743.

[20] Atia A., Adsorption of chromate and molybdate by cetylpyridinium bentonite, *Appl. Clay Sci.* 41 (2008) 73–84.

[21] Theron J., Walker J.A., Cloete T.E., Nanotechnology and Water Treatment: Applications and Emerging Opportunities.*Crit. Rev. Microbiol.* 34 (2008) 43-69.

[22] Li X.Q., Elliott D.W., Zhang W.X., Zero-Valent Iron Nanoparticles for Abatement of Environmental Pollutants: Materials and Engineering Aspects, *Crit. Rev. Solid State Mater. Sci.* 31 (2006) 111–122.

[23] He P., Zhao D.Y., Manipulating the size and dispersibility of zerovalent iron nanoparticles by use of carboxymethyl cellulose stabilizers. *Environ. Sci. Technol.* 39 (2005) 3314–3320.

[24] Ponder S.M., Darab J.G., Mallouk T.E., Remediation of Cr (VI) and Pb (II) aqueous solutions using supported, nanoscale zero-valent iron. *Environ. Sci. Technol.* 34 (2000) 2564–2569.

[25] Tang Y.B., Lee C.S., Chen Z.H., Yuan G.D., Kang Z.H., Luo L.B., Song H.S., Liu Y., He Z.B., Zhang W.J., Bello I., Lee S.T., High-quality graphenes via a facile quenching method for field-effect transistors. *Nano Lett.* 9 (2009) 1374-1377.

[26] Wang X., Zhi L., Mullen K., Transparent, conductive graphene electrodes for dye-sensitized solar cells. *Nano Lett.* 8 (2008) 323-327.

[27] Ang P.K., Chen W., Wee A.T.S., Loh K.P., Solution-gated epitaxial graphene as pH sensor. *J. Am. Chem. Soc.* 130 (2008) 14392-14393.

[28] Stoller M.D., Park S., Zhu Y., An J., Ruoff R.S., Graphene-based ultracapacitors. *Nano Lett.* 8 (2008) 3498-3502.

[29] Kim K.S., Zhao Y., Jang H., Lee S.Y., Kim J.M., Kim K.S., Ahn J.H., Kim P., Choi J.Y., Hong B.H., Large-scale pattern growth of graphene films for stretchable transparent electrodes. *Nature* 457 (2009) 706-710.

[30] Niyogi S., Bekyarova E., Itkis M.E., McWilliams J.L., Hamon M.A., Haddon R.C., Solution properties of graphite and grapheme. *J. Am. Chem. Soc.* 128 (2006)7720–7721.

[31] Zhang K., Dwivedi V., Chi C.Y., Wu J.S., Graphene oxide/ferric hydroxide composites for efficient arsenate removal from drinking water. *Journal of Hazardous Materials*.182(2010) 162–168.

[32] Deng X.J., LüL.L., Li H.W., Luo F., The adsorption properties of Pb(II) and Cd(II) on functionalized graphene prepared by electrolysis method. *J. Hazard. Mater.* 183 (2010) 923-930.

[33] Wang B., Chang Y.H., Zhi L.J., High yield production of graphene and its improved property in detecting heavy metal ions. *New Carbon Mater.* 26 (2011) 31-35.

[34] Li N., Zheng M., Chang X., Ji G., Lu H., Xue L., Pan L., Cao J., Preparation of magnetic CoFe2O4-functionalized graphene sheets via a facile hydrothermal method and their adsorption properties. *J. Solid State Chem.* 184 (2011) 953-958.

[35] Ramesha G.K., Vijayakumar A., Muralidhara H.B., Sampath S., Graphene and graphene oxide as effective adsorbents toward anionic and cationic dyes. *J. Colloid Interface Sci.* 361 (2011) 270-277.

[36] Li Y.H., Zhang P., Du Q., Peng X., Liu T., Wang Z., Xia Y., Zhang W., Wang K., Zhu H., Wu D., Adsorption of fluoride from aqueous solution by graphene. *J. Colloid Interface Sci*. 363 (2011) 348-354.

[37] Song H., Hao L., Tian Y., Wan X., Zhang L., Lv Y., Stable and Water-Dispersible Graphene Nanosheets: Sustainable Preparation, Functionalization, and High-Performance Adsorbents for Pb2+ . *Chem. Plus Chem.* 77 (2012) 379-386.

[38] Zhang K., Mao L., Zhang L. L., Chan H. S. O., Zhao X. S., Wu J., Surfactant-intercalated, chemically reduced graphene oxide for high performance supercapacitor electrodes.*J. Mater. Chem.* 21 (2011) 7302-7307.

[39] Shu Y.H., Li L.S., Zhang Q.Y., Wu H.H., Equilibrium, kinetics and thermodynamic studies for sorption of chlorobenzenes on CTMAB modified bentonite and kaolinite. *J. Hazard. Mater.*173 (2010) 47-53.

[40] Jin Y.J., Liu F., Tong M.P., Hou Y.L., Removal of arsenate by cetyltrimethylammonium bromide modified magnetic nanoparticles. *J. Hazard. Mater.* 227-228(2012) 461-468.

[41] Hummers W.S., Offeman R.E., Preparation of graphitic oxide. *J. Am. Chem.Soc.* 80 (1958) 1339.

[42] Roberts S.N.C., Williams A.C., Grimsey I.M., Booth S.W., Quantitative analysis of mannitol polymorphs. X-ray powder diffractometry—exploring preferred orientation effects *J Pharmaceut. Biomed.*28 (2002) 1149-1159.

[43] Walter W.G., Standard methods for the examination of water and wastewater, *American Journal of Public Health and the Nations Health*, 1961.

[44] Ai L.H., Zhang C.Y., Chen Z.L., Removal of methylene blue from aqueous solution by a solvothermal-synthesized graphene/magnetite composite. *J. Hazard. Mater.* 192 (2011) 1515-1524.

[45] Ma P.C., Kim J.K., Tang B.Z., Functionalization of carbon nanotubes using a silane coupling agent. *Carbon* 44 (2006) 3232-3238.

[46] Yang D., Velamakanni A., Bozoklu G., Park S., Stoller M., Piner R.D., Chemical analysis of graphene oxide films after heat and chemical treatments by X-ray photoelectron and Micro-Raman spectroscopy. *Carbon* 47 (2009) 145-152.

[47] Üstündag˘a Z., Solak A.O., EDTA modified glassy carbon electrode: Preparation and characterization. *Electrochim. Acta* 54 (2009) 6426-6432.

[48] Li D., Muller M.B., Gilje S., Kaner R.B., Wallace G.C., Processable aqueous dispersions of graphene nanosheets. *Nat. Nanotechnol.* 3 (2008) 101–105.

[49] Stankovich S., Dikin D.A., Piner R.D., Kohlhaas K.A., Kleinhammes A., Synthesis of graphene-based nanosheets via chemical reduction of exfoliated graphite oxide. *Carbon* 45 (2007) 1558-1565.

[50] Paredes J.I., Villar R.S., Martinez A.A., Tascon J.M.D., Graphene Oxide Dispersions in Organic Solvents.*Langmuir* 24 (2008) 10560-10564.

[51] Li Y. H., Du Q. J., Liu T. H., Sun J. K., Jiao Y. Q., Xia Y. Z., Xia L. H., Wang Z. H., Zhang W., Wang K.L., Zhu H.W., Wu D. H., Equilibrium, kinetic and thermodynamic studies on the adsorption of phenol onto graphene. *Mater. Res. Bull.* 47 (2012) 1898-1904.

[52] El Qada E.N., Allen S.J., Walker G.M., Adsorption of Basic Dyes onto Activated Carbon Using Microcolumns. *Ind. Eng. Chem. Res.* 45 (2006) 6044-6049.

[53]] Liu T.H., Li Y.H., Du Q.J., Sun J.K., Jiao Y.Q., Yang G.M., Wang Z.H., Xia Y.Z., Zhang W., Wang K.L., Zhu H.W., Wu D.H., Adsorption of methylene blue from aqueous solution by graphene. Colloids Surf. *B: Biointerfaces* 90(2012)197-203.

[54] Li, Y., Gao, B., Wu, T., Sun, D., Li, X., Wang, B., Lu, F., Hexavalent chromium removal from aqueous solution by adsorption on aluminum magnesium mixed hydroxide. *Water Res.* 43(2009)3067–3075.

[55] Ahmed El, N., Potential of pomegranate husk carbon for Cr(VI) removal from wastewater: Kinetic and isotherm studies. *J. Hazard. Mater.* 161(2009)132–141.

[56] El-Sikaily, A., Nemr, A. E., Khaled, A., Abdelwehab, O., Removal of toxic chromium from wastewater using green alga ulva lactuca and its activated carbon. *J. Hazard. Mater.* 148(2007) 216–228.

[57] Weckhuysen, B. M., Wachs, I. E., Surface chemistry and spectroscopy of chromium in inorganic oxides. *Chem. Rev.* 96(1996)3327–3349.

[58] Butler, J. N., *Ionic Equilibrium*, Addison-Wesley: New York, 1967.

[59] Stumm, W., Morgan, J. J., *Aquatic Chemistry: An Introduction Emphasizing Chemical Equilibria in Natural Waters*, John Wiley & Sons, Inc.: New York, 1996.

[60] Zhu J.H., Wei S.Y., Gu H.B., Rapole S.B., Wang Q., Luo Z.P., Haldolaarachchige N., Young D.P., Guo Z.H., One-Pot Synthesis of Magnetic Graphene Nanocomposites Decorated with Core@Double-shell Nanoparticles for Fast Chromium Removal. *Environ. Sci. Technol.* 46 (2012) 977-985.

[61] Leyva R.R., Jacobo A.A., Diaz-Flores P.E., Guerrero-Coronado R.M., Mendoza-Barron J., Berber-Mendoza M.S., Adsorption of chromium(VI) from an aqueous solution on a surfactant-modified zeolite. Colloids Surf. *A: Physicochem. Eng. Aspects.* 330 (2008) 35-41.

[62] Langmuir I., The Constitution and fundamental properties of solids and liquids, *J. Am. Chem. Soc.* 38 (1916) 2221–2295.

[63] Weber T.W., Chakravorti R.K., Pore and solid diffusion models for fixed-bed adsorbers, *AIChE J.* 20 (1974) 228–238.

[64] Freundlich H.M.F., Over the adsorption in solution, *Z Phys. Chem.* 57 (1906) 385–471.

[65] Sharma Y.C., Uma, Optimization of Parameters for Adsorption of Methylene Blue on a Low-Cost Activated Carbon. *J. Chem. Eng. Data* 55 (2010) 435-439.

[66] Dogan M., Alkan M., Demirbas O., Ozdemir Y., Ozmetin C., Adsorption kinetics of maxilon blue GRL onto sepiolite from aqueous solutions. *Chem. Eng. J.* 124 (2006) 89-101.

[67] Ho Y.S., Chiang C.C., Sorption Studies of Acid Dye by Mixed Sorbents. *Adsorpt. J. Int. Adsorpt. Soc* 7 (2001) 139-147.

[68] Deng X. J., Lu L. L., Li H. W., Luo F., The adsorption properties of Pb(II) and Cd(II) on functionalized graphene prepared by electrolysis method. *J. Hazard. Mater.* 183(2010)923-930.

[69] Ozer D., Dursun G., Ozer A., Methylene blue adsorption from aqueous solution by dehydrated peanut hull. *J. Hazard. Mater.* 144 (2007) 171-179.

[70] Weber W. J., Morris J. C., Kinetics of adsorption on carbon from solution. *J. San. Eng. Div.* 89 (1963) 31-36.

[71] Hameed B. H., Equilibrium and kinetic studies of methyl violet sorption by agricultural waste. *J. Hazard. Mater.* 154 (2008) 204-212.

[72] Ren T.Y., He P., Niu W.L., Wu Y.J., Ai L.H., Gou X. L., Synthesis of α-Fe2O3 nanofibers for applications in removal and recovery of Cr(VI) from wastewater. *Environ Sci Pollut Res*.20 (2013) 155-162.

[73] Fernandes A.N., Almeida C.A.P., Debacher N., Sierra M.M.S., Isotherm and thermodynamic data of adsorption of methylene blue from aqueous solution onto peat. *J. Mol. Struct.* 982 (2010) 62-65.

[74] Weng C.H., Lin Y.T., Tzeng T.W., Removal of methylene blue from aqueous solution by adsorption onto pineapple leaf powder. *J. Hazard. Mater.* 170 (2009) 417-424.

[75] Nasuha N., Hameed B.H., Adsorption of methylene blue from aqueous solution onto NaOH-modified rejected tea. *Chem. Eng. J.* 166(2011) 783-786.

In: Graphene
Editor: Bruce T. Edwards
ISBN: 978-1-62948-795-3

Chapter 3

GRAPHENE-NON-NOBLE METAL HYBRID NANOMATERIALS AS ADVANCED ELECTROCATALYSTS

***Ruizhong Zhang*[1,2] *and Wei Chen*[1,*]**
[1]State Key Laboratory of Electroanalytical Chemistry,
Changchun Institute of Applied Chemistry,
Chinese Academy of Sciences, Changchun, Jilin, China
[2]University of the Chinese Academy of Sciences, Beijing, China

ABSTRACT

Graphene, as the fundamental 2D carbon structure with exceptionally high crystal and electronic quality, has emerged as a rapidly rising star in the field of material science. Its discovery in 2004 led to an explosion of interest in the study of graphene with respect to its unique physical, chemical, and mechanical properties, opening up a new research area for multidisciplinary fields, and aiming for wide-ranging and diversified technological applications. On the other hand, due to the large surface area and high catalytic activity, nanomaterials exhibit promising applications in fuel cells as advanced electrocatalysts. In this chapter, we highlight recent advances in the development of graphene-non-noble metal nanomaterials from the standpoint of electrocatalysts. The unique properties of graphene will be discussed firstly. The exciting progress

* To whom all correspondence should be addressed. E-mail: weichen@ciac.jl.cn.

related to graphene-based nanomaterials in electrocatalysts in recent years, including graphene or its derivative (heteroatom doped graphene, etc.), and graphene-supported metal non-noble nanoparticles are summarized in detail.

Keywords: Graphene, nanomaterials; electrocatalyst, non-noble metal, fuel cells

1. Graphene and Its Properties

Graphene is a new family member of carbon materials discovered by Geim et al. in 2004. [1] Graphene is composed with only single layer of sp^2-hybrided carbon atoms densely arranged in a honeycomb crystal lattice. As a basic carbon building block, other high-dimensional graphitic materials, such as zero-dimensional fullerenes, one-dimensional nanotubes, and three-dimensional graphite can be constructed from graphene nanosheets (GNs), as shown in Figure 1. [2] Graphene has shown various electronic, thermal and optical properties, such as high carrier mobility at room temperature (~10000 $cm^2/V\ s$), [1] large theoretical specific surface area (2630 m^2/g), [3] high Young's modulus (1 TPa) [4] and excellent thermal conductivity (3000-5000 W/mK), [5] etc. Due to these unique properties, graphene has exhibited huge potential applications in different fields, including chemical sensors, electronic devices, energy conversion and storage and catalysts, and so on. [6-14] Since initially produced by mechanical cleavage from bulk graphite, great efforts have been devoted to the development of versatile and reliable physical and chemical synthetic methods for the preparation of graphene and its derivatives. Up to now, the widely used methods, including chemical vapor deposition, epitaxial growth, arc discharge approach, thermal annealing methods, etc have been successfully applied to the preparation of graphene nanosheets with high-quality, mass production and low cost [15-18].

Fuel cells have been considered promising clean energy sources with high energy density, high efficiency and low air pollution. Currently, Pt-based materials are mostly used as anode and cathode electrocatalysts for fuel oxidation and oxygen reduction. [19] However, the limited reserve in nature and high cost of Pt make it impossible to widely commercialize fuel cells. [20-22] On the other hand, catalysts are usually dispersed on a support material to improve the dispersity and the catalytic activities of catalysts, and also reduce the usage and cost of catalysts.

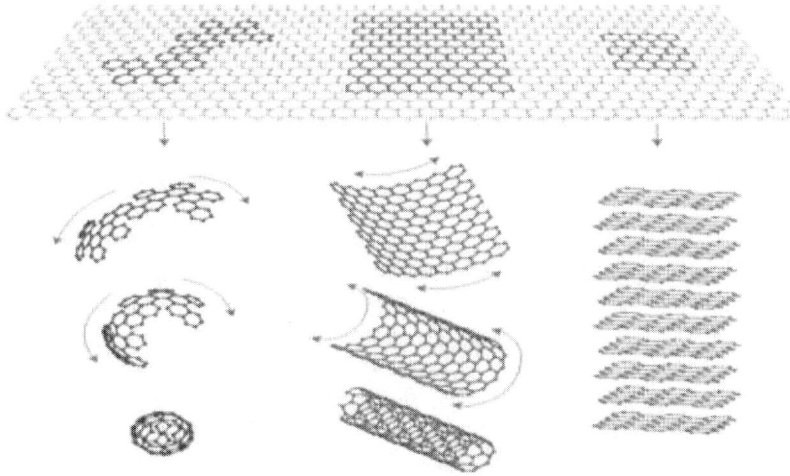

Figure 1. Carbon-based materials. Graphene can be considered a 2D building block of other carbon materials. It can be wrapped up into 0D buckyballs, rolled into 1D nanotubes or stacked into 3D graphite. Reprinted from ref. 2 with permission by Nature Publishing Group.

The state-of-the-art commercial catalysts are usually prepared by dispersing precious metal nanocatalysts on carbon supports. However, the severe corrosion and oxidation of the carbon support in the harsh operating environment would lead to the quick loss of the electrocatalytic activity of the catalysts. Based on these problems in fuel cell catalysts, to accelerate the commercialization of fuel cells, novel nanostructured electrocatalysts with high catalytic performance and low-cost, and catalyst supports with high stability have to be developed. Among the carbon materials, graphene nanosheets meet well the requirements of a catalyst support with high surface area, high electrical conductivity and high thermal stability. Moreover, it was found that heteroatom-doped graphene exhibit excellent electrocatalytic activities.

Meanwhile, abundant oxygen-containing groups on the graphene can serve as active sites for further surface functionalization with nanocatalysts to tune the catalytic properties of GNs. With these advantages, it is desirable to harness the useful properties of graphene and its derivatives in catalysts, through the incorporation with various kinds of electro-catalytically active materials. This chapter highlights the recent development of graphene-based nano-electrocatalysts for fuel cells, mainly including the heteroatom-doped graphene and graphene-supported non-noble metal nanocatalysts.

2. The Electrocatalytic Activities of Graphene and Its Derivatives

2.1. N-Doped Graphene Nanosheets (N-GNs)

From both theoretical and experimental studies, nitrogen doping is found to be an effective approach to fabricate *n*-type graphene materials and tune their properties, especially the catalytic activities. The theoretical studies illustrated that the physical and chemical properties of nitrogen-doped graphene are strongly dependent on the doping concentration and the site of the doped nitrogen atoms. Importantly, compared to pristine graphene, "activation region" can be induced by the neighbor nitrogen dopants in the nitrogen-doped graphene, which can act as catalytic sites for certain chemical/electrochemical reactions directly. It is believed that the excellent catalytic properties of the N-doped graphene nanosheets are directly related to their unique conjugated structures. The density functional theory (DFT) calculations found that the electrocatalytic activity of heteroatom-doped graphene is strongly dependent on the electron spin density and atomic charge density distribution on atoms.

According to the recent studies, the N-doped graphene sheets usually have three bonding configurations within the carbon lattice, including quaternary N (or graphitic N), pyridinic N, and pyrrolic N. [23] Among the three kinds of doped nitrogen in graphene, pyridinic N bonds with two N atoms at the edges or defects of graphene and contributes one p electron to the π system of graphene. Pyrrolic N refers to the N atoms that contribute two p electrons to the π system, although unnecessary bond into the five-membered ring, as in pyrrole. [24] Quaternary N refers to the N atoms that substitute for C atoms in the hexagonal ring. The pyridinic N and quaternary N are sp^2 hybridized and pyrrolic N is sp^3 hybridized. Except for the three identified nitrogen types, oxides of pyridinic nitrogen have also been reported in nitrogen-doped graphene, [25] where the nitrogen atom bonds with two carbon atoms and one oxygen atom.

Moreover, nitrogen doping can lead to much change of the electronic structure and energy levels of graphene. For instance, the Fermi level of N-doped graphene shifts above the Dirac point, [26] the density of state near the Fermi level is suppressed, [27] and thus the band gap between the conduction band and the valence band will be opened. Therefore, in comparison with pristine graphene, the N-doped graphene materials are a type of promising non-precious-metal catalysts for fuel cells, which could be the alternative for the Pt-based electrocatalysts. In addition to the low cost, fuel cell catalysts based on N-doped graphene exhibit enhanced catalytic activity and durability. Meanwhile, N-doped graphene well facilitate the efficient four-electron oxygen reduction reaction because of their unique electronic structures.

Nitrogen can be doped into graphene during the synthesis of graphene or by post treatment of the pre-synthesized graphene sheets with the presence of N-containing precursors. Among the direct synthesis methods, chemical vapor deposition (CVD) has been widely used to prepare high-quality graphene and N-doped graphene sheets. In the reported CVD routs, gaseous or liquid carbon- and nitrogen-containing species, such as NH_3, acetonitrile, pyridine etc. can be used as carbon and nitrogen sources. [28-30] At high temperature, the precursors will dissociate and produce N-doped graphene sheets through precipitation on the surface of metal catalysts. By CVD method, the doping concentration and the bonding configuration of nitrogen can be effectively tuned by controlling the experimental conditions, such as temperature, gas flow rate, the amounts of the carbon and nitrogen precursors, and metal catalysts etc. For instance, Wei et al. [28] synthesized N-doped graphene through CVD method by using copper film on a Si substrate as the catalyst, CH_4 and NH_3 as the C and N sources, respectively. From the XPS spectra shown in Figure 2, nitrogen has been successfully doped in the graphene sheets. Also from the N 1s spectrum (Figure 2c), the doped nitrogen atoms exhibit three different bonding characters, that is pyridinic N at 398.2 eV, pyrrolic N at 400.1 eV, and graphitic N at 401.7 eV. Moreover, it was found that the N content in graphene decreased from 8.9 to 3.2 and 1.2 at% when the ratio of NH_3/CH_4 changed from 1:1 to 1:2 and 1:4. The resulting N-doped graphene behaves like an n-type semiconductor.

It can be seen that in the CVD methods, metal substrates are used as catalysts for graphene or N-doped graphene growth. From the viewpoint of catalysis, the possible metal residue in the graphene products may influence their catalytic activities. Therefore, except for CVD, other methods, such as segregation growth approach (SGA), solvothermal approach (STA), arc-discharge approach (ADA), plasma treatment approach (PTA), and thermal treatment approach (TTA) have also been developed to synthesize N-doped graphene. [31-34] In the previous studies, it has been found that nitrogen may be more easily doped at the defects and edge of graphene.

Recent reports demonstrated that N-GNs show excellent electrocatalytic activity toward the oxygen reduction reaction (ORR) in alkaline medium [34-37].

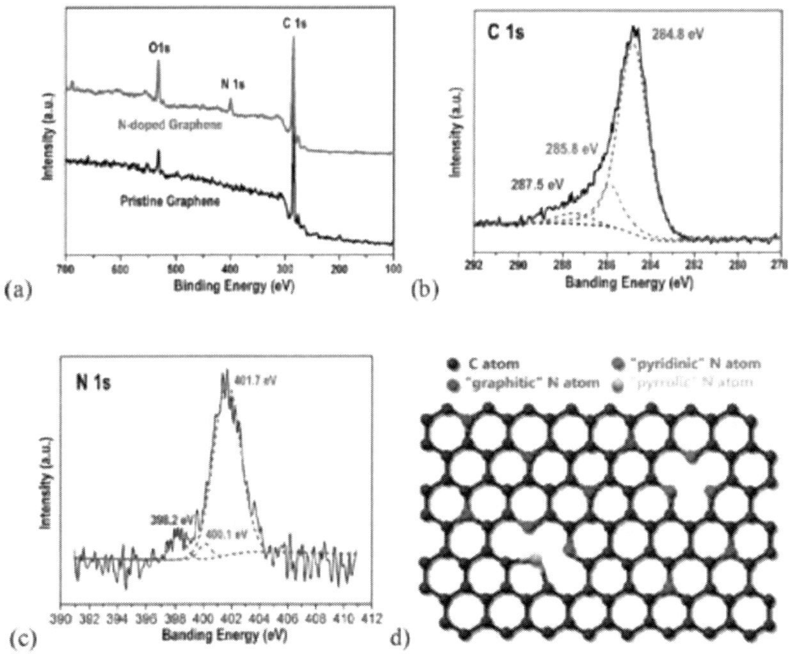

Figure 2. (a) XPS survey spectra of the pristine graphene and N-doped graphene. (b, c) XPS spectra of C 1s (b) and N 1s of the N-doped graphene. (d) Schematic representation of the N-doped graphene. Reprinted from ref. 28 with permission by the American Chemical Society.

Ruoff and co-workers [37] studied the catalytic activity of N-doped GNs for ORR and found that the bonding state of the N atom has a significant effect on the selectivity and catalytic activity for ORR. The electrochemical results showed that the graphitic N content may determine the limiting current density of ORR, while the pyridinic N content can improve the onset potential for ORR ($4e^-$ dominated reduction process). However, the total N content in the N-GNs catalyst does not play an important role in the ORR process. In another work, Xia group [34] proposed a facile, catalyst-free thermal annealing approach for large-scale synthesis of N-GNs using low-cost industrial material melamine as the nitrogen source. The as-synthesized N-GNs mainly contain pyridine-like nitrogen atoms and exhibited excellent electrocatalytic activity toward ORR. The cyclic voltammetry (CV) and rotating ring-disk electrode (RRDE) techniques demonstrated that the N-GNs have more positive onset potential and much larger reduction current of ORR in comparison with the pristine graphene, which could be ascribed to the faster reaction kinetics with a higher transferred electron number of ORR at the N-GNs-modified electrode surface. Moreover, such N-GNs can catalyze the ORR in a one-step four-electron pathway with excellent catalytic activity, selectivity and stability. The high catalytic activity of N-GNs could be attributed to the changed electronic structure of graphene sheets upon the incorporation of nitrogen atoms, which favors the reactivity of the neighborly linked carbon atoms. Although carbon nitride (CN) materials prepared by pyrolyzing the industrial available melamine (or cyanamide etc.) may serve as a potential metal-free catalyst for ORR, the low electrical conductivity of CN materials is one of the major obstacles for their application in fuel cells. In this regard, Yang et al. [35] fabricated graphene-based carbon nitride (G-CN) nanosheets by a nanocasting technique.

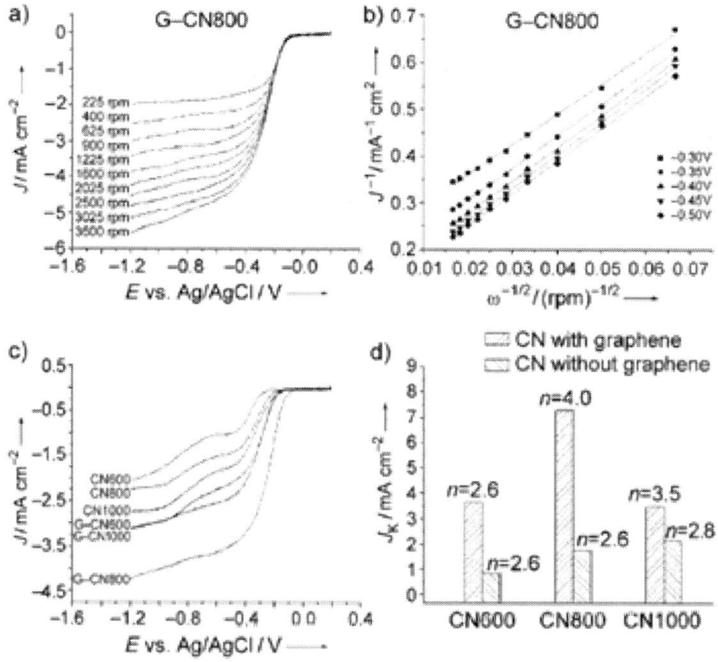

Figure 3. (a) Rotating disk electrode (RDE) linear sweep voltammograms of graphene-based carbon nitride (G-CN) annealed at 800 °C (G-CN800) in O_2-saturated 0.1 M KOH with various rotation rates at a scan rate of 5 mV/s. (b) Koutecky-Levich plots of G-CN800. (c) RDE voltammograms of G-CN and CN nanosheets at a rotation rate of 1600 rpm. (d) Electrochemical activity given as the kinetic-limiting current density (J_k) at -0.40 V for all G-CN and CN nanosheets. Reprinted from ref. 35 with permission by Wiley-VCH.

The obtained G-CNs materials not only possess high nitrogen content, thin thickness, high surface areas, and large aspect ratios, but also show enhanced electrical conductivity. The authors found that the pyrolysis temperature can change the amount of N-bonding configurations. By comparing the rotating disk electrode (RDE) voltammograms (Figure 3c), the onset potentials of all the G-CNs nanosheets are more positive than those of CN nanosheets without graphene, and the G-CN nanosheets annealed at 800 ^{o}C (G-CN800) exhibits the most positive onset potential. From Figure 3, the G-CN800 nansheets exhibit a one-step four-electron transfer pathway with a high kinetic current density of 7.3 mA/cm^2. Also from the comparison shown in Figure 3d, the CN nanosheets without graphene typically show a two-electron transfer pathway and low current densities for ORR. It was proposed that the pyridinic nitrogen atoms play crucial role in determining the catalytic activities of the G-CN nanosheets. Due to the strong electron-accepting ability, pyridinic nitrogen can induce a net positive charge on the adjacent carbon atoms, which is favorable for the ORR.

From the above results, N-doped graphene nanosheets exhibit much different catalytic properties from those of pristine graphene. N-GNs exhibit excellent ORR electrocatalytic activities, long-term operation stability, and high tolerance to methanol in alkaline fuel cells, which shows the promising possibility to replace expensive platinum-based catalysts.

2.2. B-Doped Graphene Nanosheets (B-GNs)

Chemical doping of GNs with foreign boron (B) atoms is another effective way to modify the electronic properties, the surface chemistry of GNs, and consequently to tune the electrocatalytic properties. [38] Theoretical studies have demonstrated that after doping in graphene network, boron is bonded to the sp^2 carbon, which can modulate the band structure of graphene. Actually calculations indicated the B-doped graphene exhibits high conductivity and its Fermi level shifts 0.65 eV below the Dirac point. [39] As expected, B-doped graphene exhibits p-type behavior. [40, 41]

Similar to the N-GNs, B-doped GNs can also be prepared through various methods, such as arc-discharge approach, chemical vapor deposition (CVD), thermal-annealing, solvothermal method and the Wurtz-type reductive coupling (WRC) strategy and so on. Rao et al. [33] reported the synthesis of B-GN by using arc discharge of graphite electrode in the presence of B_2H_6 vapor. The synthesized B-GNs contained two to three layers with B content up to 3.1 at.% and exhibited higher electrical conductivity than the non-doped graphene. Wang group [42] prepared B-GNs via a solvothermal method using tetrachloromethane (CCl_4) and boron tribromide (BBr_3) as the raw materials, and an alkali metal (K) as a reductant. The content of boron in the product was determined to be 1.1 at.% and the average size of the as-made B-GNs is around 35 nm. In another work, Li and co-workers [43] successfully synthesized B-GNs through CVD process by using copper foil as substrate and boron power instead of the usually used boron hydride as the boron source. It was suggested that the B doping has little influence on the graphene surface morphology and the produced B-GNs still maintained a layered structure with predominantly three to five layers. In the above-mentioned methods, special instruments are required, and especially the transition metal substrates used in the CVD methods may degrade the electronic properties of the products and further limit their application as catalysts for fuel cells. In this regard, Lu et al. [44] demonstrated a Wurtz-type reductive coupling (WRC) reaction as a new bottom-up method for the rapid preparation of high-quality B-GNs using boron tribromide (BBr_3) as the boron sources. Such method can avoid the by-products to ensure the formation of pure B-GNs. In the WRC method, the B content was determined to be 1.02 at.% and it can be easily tuned by adjusting the amount of added BBr_3. In addition, this boron doping can bring more holes to the valence band of graphene, and increase the carrier concentration of B-GNs, thus enhancing its conductivity. By a catalyst-free thermal annealing approach, Sheng et al. [45] successfully realize the B-doping into GN frameworks in the presence of boron oxide (B_2O_3). During the thermal annealing, the removal of oxygen species and defects in the graphene framework at high temperature can provide active sites for boron doping. With such approach, the obtained B-GNs display a flake-like structure similar to pristine graphene with an average thickness of *ca.* 2 nm, and the B content up to 3.2 at.%. Meanwhile, the B-C bonds were formed in the annealing process characterized by the X-ray photoelectron spectroscopy (XPS) and the Shirley algorithm.

Due to the modified electronic band structure, B-doped graphene exhibits higher electrocatalytic activity than pristine graphene. Xia group [45] studied the electrocatalytic activity of B-GNs for ORR. From the CV and RDE measurements, B-GNs show a positive shift of onset potential and enhanced reduction current for ORR compared to pristine graphene, revealing the higher electrocatalytic activity of B-GNs towards ORR. From the voltammetric curves of RDE and the Koutechy–Levich plots, ORR on B-GNs proceeds with an efficient one-step four-electron pathway. Therefore, the enhanced catalytic property of B-GNs could be due to the fast reaction kinetics with a high transferred electron number per oxygen molecule. Moreover, the synergetic effect of N and B co-doping in graphene can further improve the catalytic activity of graphene-based materials. [38, 46] For instance, Qiao group [38] synthesized B- and N-codoped graphene sheets by a two-step strategy. The reported method not only enables the incorporation of heteroatom at selected sites of the graphene framework to induce a synergistic enhancement of the activity, but also prevents the formation of inactive by-products. The co-doped graphene catalyst exhibited higher catalytic activity for the ORR compared to the singly N- or B-doped graphene. The B, N-GNs demonstrated a four-electron ORR pathway in an alkaline medium, complete methanol tolerance and excellent long-term stability, which are superior to those for the commercial Pt/C catalyst. Overall, the B-doping or B, N co-doping into graphene networks can tailor the electronic structures of graphene materials and consequently improve their electrocatalytic performance towards ORR, which make them promising metal-free cathode catalysts in fuel cells.

2.3. Other Heteroatom-Doped Graphene Nanosheets

Except for the N-GN and B-GN for ORR, other hctcroatoms (such as S, I, P, Se)-doped or dual doped graphene materials have also been investigated as metal-free electrocatalysts. [36] Theoretical studies using simulation calculations indicated that breaking the electroneutrality of graphitic materials to create charged sites is a key factor in enhancing the ORR catalytic activity. Phosphorus (electronegativity: 2.19) has a smaller eletronegativity to carbon (electronegativity: 2.55), sulfur (electronegativity: 2.58) and selenium (electronegativity: 2.55) have the close electronegativity to carbon. Therefore, changing the electronic arrangement of graphene by doping P, S, Se could be a practical strategy to significantly improve the ORR electrocatalytic activity of graphene. [47] Hou and co-workers [48] reported a facile, low-cost and scalable thermal annealing method for the synthesis of P-GNs using graphite oxide (GO) and triphenylphosphine (TPP) as carbon and phosphrous sources, respectively.

The CV recorded at the P-GNs (Figure 4a) shows the obvious catalytic activity of the P-GNs for ORR. From the rotating disk voltammograms shown in Figure 4b, the ORR current density obtained from the P-GN is much higher than that from GNs without P doping over the entire potential range. More importantly, from the inset of Figure 4b, the P-GNs displayed a significantly increased ORR onset potential, which is much more positive than that of GNs and close to that of the commercial Pt/C. More importantly, the P-GNs exhibited remarkable tolerance to methanol crossover effect (Figure 4c) and excellent long-term stability (Figure 4d) compared with the commercial Pt/C catalyst, which endow P-GNs great potential to be efficient metal-free electrocatalysts in alkaline fuel cells.

Yang et al. [49] demonstrated the successful fabrication of S-GNs and Se-GNs by directly annealing graphene oxide (GO) and the benzyldisulfide (BDS) or diphenyl diselenide in argon. The electrocatalytic studies showed that the S-GNs and the Se-GNs have excellent catalytic activity, long-term stability, and high methanol tolerance in alkaline media for ORR. The results also showed that graphene doped with S or Se, which have the similar electronegativity with carbon, exhibit better catalytic activity than the commercial Pt/C in alkaline medium, indicating the potential application of these doped graphene materials in fuel cells as metal-free catalysts.

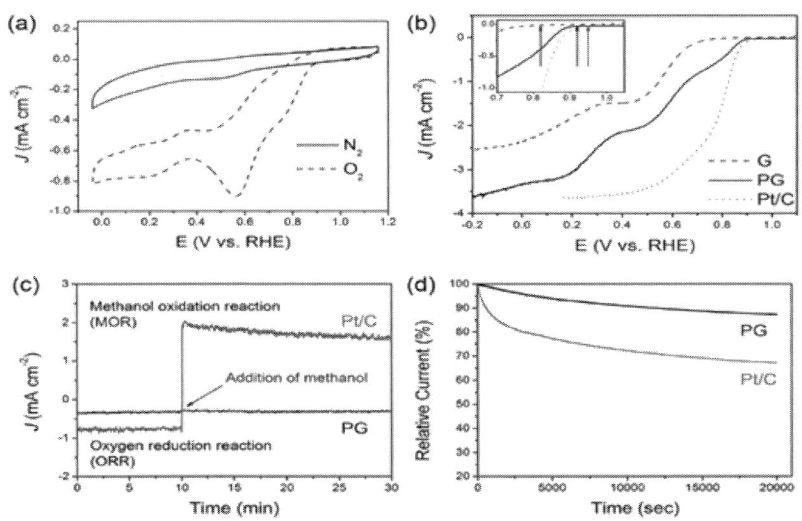

Figure 4. (a) CVs of PG in a N_2- and O_2-saturated 0.1M KOH solution at a scan rate of 100 mV/s. (b) LSV curves of graphene (G), PG and Pt/C in an O_2-saturated 0.1 M KOH solution at a scan rate of 10 mV/s and a rotation speed of 1600 rpm. Inset shows the onset potential of G, PG and Pt/C. (c) Methanol crossover effect tests of PG and Pt/C upon addition of 3 M methanol after about 10 min in an O_2-saturated 0.1 M KOH solution at 0.7 V. (d) Stability evaluation of PG and Pt/C for 20000 s in an O_2-saturated 0.1 M KOH solution at 0.7 V and a rotation speed of 1600 rpm. Reprinted from ref. 48 with permission by Wiley-VCH.

In addition, iodine-doped graphene (I-GN) has also been successfully fabricated by annealing GO and iodine by Huang group. [50] The authors found that the I-GN has the close onset potential with Pt/C and higher current density than the Pt/C catalyst for ORR in alkaline medium. However, the simple physical accumulation of molecule iodine in the doped graphene surface is not helpful for enhancing the ORR activity of graphene. It is believed that the formation of the I_3^- structure play a crucial role for the ORR activity enhancement.

3. The Electrocatalytic Activities of Graphene-Non-Noble Metal Hybrids

A major limiting factor of energy-conversion efficiency for present fuel cells is the sluggish kinetics of fuels oxidation at anode and the ORR at cathode. Because of the large surface area, excellent conductivity, and good chemical stability, graphene is a promising support material for metal electrocatalysts.

Many studies have demonstrated that nanoparticle catalysts supported on graphene surface show enhanced catalytic activities. The high catalytic activities of graphen-metal hybrids may be ascribed to the charge transfer across the graphene-metal interface originated from the GN-metal spacing and the Fermi level difference between the GNs and the metal. [51] We recently successfully synthesized 2 nm Ir_xV bimetallic nanoclusters assembled on reduced graphene nanosheets (Ir_xV/rGO) by a simple "surfactant-free" solution phase process [52].

We found the Ir_xV/rGO hybrids exhibited composition-dependent catalytic activity for ORR and the Ir_2V/rGO clusters displayed the highest catalytic performance. Moreover, compared to commercial Pt/C catalyst, larger reduction current density and much higher tolerance to methanol crossover were obtained on the Ir_2V/rGO, and the RDE and RRDE experiments demonstrated that ORR on the Ir_2V/rGO exhibits the most efficient four-electron pathway. It is anticipated that the Ir_xV nanoclusters assembled on rGO can be used as promising Pt-free cathode in alkaline fuel cells. It should be noted that the high cost and scarcity of noble metals hindered severely their widespread application and the commercialization of fuel cells. Consequently, alternative non-noble metal electrocatalysts supported on graphene have been actively pursued.

3.1. GNs-Supported Cobalt Hybrid Materials as Electrocatalysts

Graphene nanosheets-supported cobalt oxides or sulfides are usually prepared either through in situ chemical deposition/direct self-assembly in solution with the presence of GO and metal precursors or post-treatment methods. Several methods have been reported to prepare graphene supported cobalt-based electrocatalysts, including solution-phase reaction and hydrothermal approach, solution-phase reaction and thermal annealing approach, solution-phase self-assembly approach and electrodeposition. For example, Wang et al. [53] synthesized cobalt sulfide nanopartices supported on reduced graphene oxide (rGO) sheets by a low-temperature solution-phase reaction followed by a high-temperature annealing step. With this method, cobalt sulfide nanoparticles were selectively grown on rGO sheets without free growth of unattached nanoparticles in solution. The cobalt sulfide nanoparticles on rGO have an average particle size of about 10-20 nm and the $Co_{1-x}S$ phase with hexagonal structure.

Recent studies showed that Co_3O_4 nanoparticles supported on graphene have excellent catalytic activities for ORR. Dai and co-workers [54] synthesized 12-25 nm Co_3O_4 nanoparticles supported on reduced mildly oxidized graphene oxide (rmGO) by a solution-phase reaction and hydrothermal method.

Meanwhile, when NH_4OH was added in the synthesis steps of Co_3O_4/rmG, the hydrolysis of Co^{2+} and its oxidation can be mediated, and N-doped hybrid material (Co_3O_4/N-rmG) can be produced. The Co_3O_4 nanoparticles in the Co_3O_4/N-rmG is smaller (~4-8 nm) than those in the Co_3O_4/rmG, which can be attributed to the NH_3 coordination with cobalt cations.

The CV measurements revealed that Co_3O_4 nanocrystals or rmGO alone exhibited very poor ORR activity, while the Co_3O_4/rmGO hybrid showed a much more positive ORR onset potential and higher cathodic currents, suggesting the synergistic ORR catalytic activity of Co_3O_4 and rmGO in the hybrid.

As shown in Figure 5a, the catalytic activity can be greatly enhanced with nitrogen doped in the graphene sheets. The CV of ORR on the Co_3O_4/N-rmGO hybrid exhibited a more positive peak potential and higher peak current compared to that of the Co_3O_4/rmGO. The rotating-disk voltammograms of ORR recorded at the Co_3O_4/N-rmGO (Figure 5c) revealed $4e^-$ transfer in the potential range of 0.6-0.75 V, and a half-wave potential of 0.83 V at 1600 rpm, which is similar to that of Pt/C, and more positive than that of the

Co_3O_4/rmGO (Figure 5b). More importantly, the much smaller Tafel slope of 42 mV/decade at low over-potentials (Figure 5d) than the Co_3O_4/rmGO hybrid further indicates the excellent ORR activity of the Co_3O_4/N-rmGO hybrid catalyst.

The X-ray absorption near edge structure (XANES) measurements showed the possible formation of interfacial Co-O-C and Co-N-C bonds in the Co_3O_4/N-rmG hybrids, a higher electron density at O site and a lower electron density at the Co site. These special bond structures between Co_3O_4 and N-rmGO, and changes in the chemical bonding environment for C, O and Co atoms in the hybrids material were proposed to be responsible for the synergistic ORR catalytic activity.

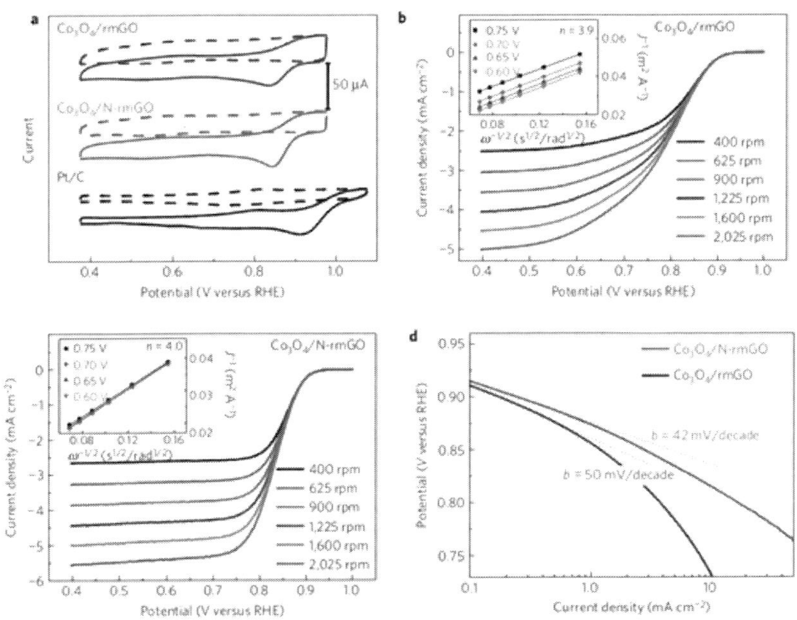

Figure 5. (a) CV curves of Co_3O_4/rmGO hybrid, Co_3O_4/N-rmGO hybrid and Pt/C on glassy carbon electrodes in O_2-saturated (solid line) or Ar-saturated 0.1 M KOH (dash line). Catalyst loading is 0.17 mg/cm^2 for all samples. (b) Rotating-disk voltammograms of Co_3O_4/rmGO hybrid and (c) Co_3O_4/N-rmGO hybrid in O_2-saturated 0.1 M KOH with a sweep rate of 5 mV/s at the different rotation rates indicated. The insets in b and c show the corresponding Koutecky-Levich plots (j^{-1} versus $\omega^{-0.5}$) at different potentials. (d) Tafel plots of Co_3O_4/rmGO and Co_3O_4/N-rmGO hybrids derived by the mass-transport correction of the corresponding RDE data. Reprinted from ref. 54 with permission by Nature Publishing Group.

It is well documented that the catalytic properties of metal-based nanoparticles are strongly dependent on their size, shape and composition. Guo et al. [55] recently prepared monodisperse Co/CoO nanoparticles deposited on GNs surface with desired morphology and composition through a solution-phase self-assembly method. Importantly, in the study the authors found that both the GN-Co/CoO interaction and the dimension of Co and CoO are the important factors for the high catalytic performance. From the CVs of ORR, the GN-Co/CoO hybrid showed a much stronger reduction current peak than GN and carbon-supported Co/CoO nanoparticles. ORR polarization curves from GN-Co/CoO exhibited a sharp increase and showed a more positive half-wave potential than C-Co/CoO. The Koutecky-Levich plots showed the most efficient four-electron ORR process on the GN-Co/CoO, indicating the catalytic enhancement from GNs support. Furthermore, the ORR activities of the GN-Co/CoO were found to depend on the CoO thickness, with the optimal thickness of 1 nm. Compared to the commercial Pt/C, the GN-Co/CoO NPs have a steeper polarization curve with larger current density, and longer-term stability.

On the other hand, it has been found that graphene nanosheets functionalized with transition metal macrocycle compounds can serve as good alternatives to Pt-based catalysts. For example, Jeon et al. [56] prepared cobalt [tetrekis(*o*-aminophenyl)porphyrin] (CoTAPP) covalently anchored onto the surface of three types of carbon nanomaterials (graphene, single-walled carbon nanotubes, and multi-walled carbon nanotubes) via diazonium salt reactions, and compared their electrocatalytic activity for ORR. From the electrochemical studies, GNs anchored with CoTAPP showed the most efficient electrocatalytic performance for ORR among the carbon-based materials. Recently, Dai and co-workers [53, 54] also found the $Co_{1-x}S$/rGO hybrid is also an excellent ORR catalyst showing higher ORR current density than any other reported cobalt chalcogenide-based catalysts in acidic media. The hybrid gave an ORR onset potential at about 0.8 V versus RHE and a current density of 1.1 mA/cm^2 at 0.7 V. The good catalytic performance was attributed to the intimate coupling between the $Co_{1-x}S$ nanopartiales and rGO, due to the direct nucleation and growth approach. Unlike the free $Co_{1-x}S$ particles and their physical mixture with rGO, the $Co_{1-x}S$ particles in $Co_{1-x}S$/rGO hybrid are electrically wired to the rGO sheets forming a conducting network, which facilitates the ORR through facile charge transport between oxygen molecules adsorbed on $Co_{1-x}S$ active sites and the electrode. This study further revealed that direct growth of nanopartices on rGO can improve the catalytic performance.

3.2. GNs-Supported Iron Hybrid Materials as Electrocatalysts

Up to now, different transition metal macrocyclic compounds supported on carbon materials have been widely investigated for ORR. [57-59] Moreover, nitrogen doping can not only enhance the conductivity of graphene, but also make the carbon atoms around nitrogen atoms active for ORR. GNs-supported iron/nitrogen hybrids (GN/Fe-N) have attracted increasing attention due to their promising catalytic activity towards the ORR, along with the utilization of abundant, inexpensive precursor materials. Generally, the GN/Fe-N hybrid catalysts can be synthesized by treating precursors containing nitrogen, Fe salts adsorbed on graphene or functionalized graphene nanomaterials under optimized conditions. Recently, Jahan et al. [60] found that functionalized GNs can be used as a building block in the assembly of iron porphyrin framework (MOF), which exhibited enhanced catalytic activity for ORR. In the reported method, GO and rGO were first synthesized, the rGO sheets were then functionalized with donor-π-acceptor dye (G-dye). After the functionalization of the G-dye with iron-porphyrin (MOF), the G-dye-MOF hybrid can be obtained. Except for iron-porphyrin, iron phthalocyanine (FePc) [61] and Fe-Pentaethylenehexamine [62] have also been assembled on functionalized graphene as novel and efficient non-precious catalysts for the ORR. For example, Hou group [63] developed a facile method based on liquid-phase to prepare FePc/GX (GX = NG, GO and rGO) hybrids. In the study, the FePc with a typical columnar morphology with a length of about 100 nm were produced on the N-doped graphene surface. These results unambiguously showed that FePc can be well anchored on graphene surface via π-π stacking. Moreover, the much closer combination between FePc and graphene substrate compared to FePc/CNTs can make contribution to their excellent electrocatalytic performance for ORR.

Except for the Fe-macrocycle precursors, it has been found that iron-based catalysts supported on N-doped graphene can also be synthesized by chemical impregnation of simple iron sources ($FeCl_3$, $Fe(Ac)_2$, etc.), nitrogen-containing precursors and GO under thermal treatment. [64-66] For instance, Byon et al. [67] developed a facile method to synthesize the Fe-N-rGO catalyst. In their experiment, GO was first prepared by a modified Hummers method and the g-C_3N_4 precursors were thermally polymerized from dicyandiamide and employed as the nitrogen source. GO was then added to the suspension of $FeCl_3$ and g-C_3N_4, and the mixture was reduced in NH_3/N_2H_4 solution at 80-100 °C. The resulting powder mixture ($FeCl_3$/g-C_3N_4/rGO) was collected, dried and heated at 800 °C for 2 hours under Ar. Finally the pyrolyzed product was acid-leached in 2 M H_2SO_4 at 80 °C for 3 h to remove extraneous Fe species. The EXAFS analysis showed that the Fe-N configurations in the Fe-N-rGO sample may be a combination of Fe-N_{2+2} and /or its incomplete form (Fe-N_3). Recently, Tsai and co-workers [64] also synthesized the nitrogen-doped graphene-supported carbon-containing iron nitride (FeCN/NG) via thermal treatment of N-doped graphene (NG) and N-coordinated Fe complex.

Wu et al. [68] reported an interesting work on the controllable assembly of non-precious iron oxide nanoparticles on three-dimensional (3D) graphene networks as ORR catalysts. In the strategy as demonstrated in Figure 6, graphene oxide (GO) was first dispersed in water by sonication, reaching a concentration up to 1.5 mg/mL. Iron acetate and polypyrrole (PPy) were then slowly added to the GO dispersion to form a stable aqueous suspension. Subsequently, these ternary components were hydrothermally assembled at 180 °C for 12 h to form a graphene-based 3D hydrogel. The as-prepared hydrogel was directly dehydrated via a freeze-drying process to maintain the 3D monolithic architecture and then heated at 600 °C for 3h under nitrogen. The synthesized interconnected, porous 3D graphene framework with continuous macropores can not only provide large surface area for the dispersion of Fe_3O_4 nanoparticles, but also enhance the interface contact to suppress the dissolution and agglomeration of nanoparticles.

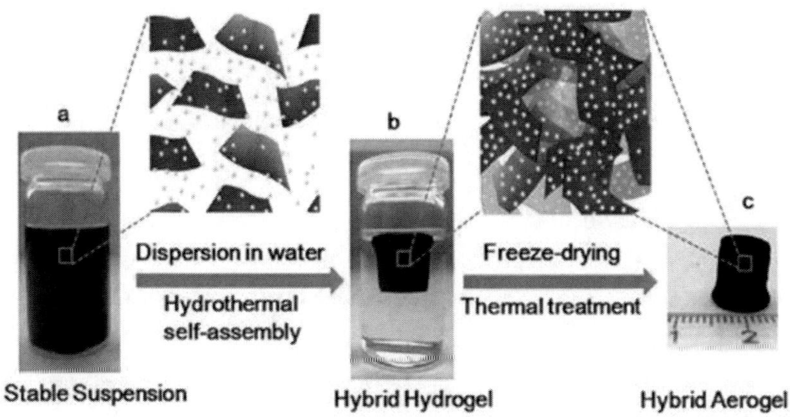

Figure 6. Fabrication process of the 3D Fe_3O_4/N-GAs catalyst. (a) Stable suspension of GO, iron ions, and PPy dispersed in a vial. (b) Fe- and PPy-supporting graphene hybrid hydrogel prepared by hydrothermal self-assembly and floating on water in a vial, and its ideal assembled model. (c) Monolithic Fe_3O_4/N-GAs hybrid aerogel obtained after freeze-drying and thermal treatment. Reprinted from ref. 68 with permission by the American Chemical Society.

The abundant Fe-N-C active sites at the Fe_3O_4 NP interface and the pyrrolic and pyridinic N active sites make the prepared 3D materials have high electrocatalytic activity for ORR.

The graphene-supported iron-based hybrids have been extensively examined as active catalysts for ORR. Interestingly, Parvez and co-workers [66] recently reported that iron (Fe) nanoparticles incorporated into N-GNs exhibited an outstanding electrocatalytic performance for ORR, which was attributed to the synergistic coupling between Fe and N-GNs. In the report, the N-GNs supported with 5 wt% Fe nanoparticles showed almost 2 times higher current density than Pt/C and displayed an excellent methanol crossover effect in alkaline solution. Moreover, the Fe-incorporated N-GN showed almost four-electron transfer ORR process and superior stability in both alkaline and acidic solution, which outperformed the platinum and N-GN catalysts. Zhang et al. [63] prepared a series of iron phthalocyanine (FePc) and graphene composites, and studied the ORR activity and stability of FePc supported on three kinds of graphene derivatives (graphene oxide (GO), reduced GO (rGO) and nitrogen-doped graphene (NG)). From the electrochemical results, the FePc/NG composites showed the highest ORR catalytic activity with four-electron reduction process, which is even superior to that of commercial Pt/C catalyst. The improved catalytic activity was attributed to the much lower overall resistance of FePc/NG and the enhanced electrical conductivity confirmed by electrochemical impedance spectra (EIS). In another work, Byon and co-workers [67] reported the synthesis of Fe-N-rGO catalyst with high ORR activity and stability in acid medium. The excellent electrocatalytic activity may be ascribed to the active sites of a combination of $Fe\text{-}N_{2+2}$ and/or its incomplete form ($Fe\text{-}N_3$), which represent the catalytically different sites from the $Fe\text{-}N_4$ group in macrocycles.

3.3. Other GN-Supported Non-Noble-Metal Hybrid Materials as Electrocatalysts

Among non-noble metals, except for iron and cobalt, nickel (Ni) is another versatile catalytic material. Ni can be easily converted to $Ni(OH)_2$, and the Ni^{2+}/Ni^{3+} redox centers show high catalytic activity toward oxidation of ethanol in alkaline media. [69, 70] However, the poor electrical conductivity of both nickel hydroxide and oxides largely lower the catalytic performance to some extent. Recently, graphene has been used as an effective support material to improve the electrocatalytic activities of nickel particles for methanol and ethanol oxidation. For instance, Zhang and co-workers [71] prepared 10-20 nm Ni nanoparticles supported on GNs by reducing GO and $NiCl_2$ with hydrazine. The composition, size and dispersion of Ni nanoparticles as well as the oxygen content of graphene sheets could be roughly controlled by varying the heating temperature. The study showed that the Ni complex annealed at 500^0C to form Ni/G-500 catalyst has the highest activity and stability for methanol electro-oxidation in an alkaline media with peak current density of 20 mA/cm^2. Graphene-supported nickel oxide composite was also synthesized by Ghamdi et al. [72] as electrocatalyst for methanol oxidation through a simple chemical reduction of the mixture of GO and $Ni(NO_3)_2$ salt. Many studied have illustrated that bimetallic alloys usually have higher catalytic activities than monometallic catalysts. Wang and coworkers [73] synthesized nickel and cobalt bimetallic nanoparticles on graphene (Ni-Co/G) via a simple solution process and studied the electrocatalytic properties of the synthesized Ni-Co/G and Ni-G for ethanol oxidation. Compared with Ni/G catalyst, the Ni-Co/G showed higher electroactivity and better stability for ethanol oxidation.

To develop highly efficient and low-cost catalysts for ORR, manganese oxides are particularly attractive candidates due to their high catalytic activity, high abundance, low cost, and lack of environmental issues. [74, 75] Similar to nickel oxides, the intrinsically low electrical conductivity of manganese oxides limits their application in electrocatalysts. By taking advantage of the high chemical stability, high conductivity and the high specific surface area of graphene, manganese oxides/graphene hybrids can enhance the electrocatalytic activities. Yang et al. [76] prepared MnO_2-graphene composite by mixing the aqueous graphene nanosheet colloids and aqueous MnO_2 dispersions and the hybrid material demonstrated good catalytic performance for ORR. Interestingly, Lee and co-workers [77] reported ion-liquid (IL) moiety modified reduced graphene oxide (rGO) with the anchoring of manganese oxide (Mn_3O_4). The IL moiety in the composites can increase not only the conductivity of the material, but also the electrocatalytic activity. With the optical amount of Mn_3O_4 nanoparticle content (19.2 %), the hybrid exhibited similar catalytic activity to a Pt/C electrode via a one-step, quasi-4-electron ORR pathway. In a recent report, Cui and co-workers [78] found that through substitution of Co^{2+} by Mn^{3+}, the resulting $MnCo_2O_4$/GN hybrid showed much enhanced ORR activity over the Co_3O_4/GN.

In addition, it is known that Cu_2O can easily react with oxygen in wet air to form CuO, indicating that Cu_2O can donate electrons to oxygen. Based on this, Yan et al. [79] recently reported the synthesis of Cu_2O (~4 nm) nanoparticles dispersed on reduced graphene oxide (Cu_2O/rGO) by reducing copper acetate supported on graphite oxide with diethylene glycol as both solvent and reducing agent. In the study, the authors found that the Cu_2O nanoparticles can easily get electrons through graphene sheets and then donate electrons to oxygen, showing high ORR catalytic activity. The Cu_2O /rGO composite also showed excellent tolerance to CO and methanol, making it a potential noble metal-free cathode catalyst in fuel cells.

CONCLUSION

Graphene, a single layer two-dimensional carbon structure, has attracted an explosive interest in recent years. Among the carbon-based materials, graphene offers the advantages of unique physical, chemical and mechanical properties, which may open up a new research area for materials science. Due to the high chemical stability, high conductivity and high specific surface area, graphene is a kind of ideal catalyst support for fuel cells. This chapter summarizes the recent advances of graphene-based materials from the standpoint of non-noble-metal electrocatalysts. Owing to the principally advantageous electronic and electrochemical properties, graphene has been explored extensively either as metal-free catalyst through molecule engineering such as heteroatom doping or as a unique catalyst support for fabricating functional non-noble-metal catalysts. Promising results showed that the unique properties of graphene and the synergistic interaction between graphene and the supported components are very favorable for enhancing the electrocatalytic performance for oxygen reduction, methanol and ethanol oxidation reactions. Through further structure and composition optimization, graphene-based catalysts may serve as promising alternatives to Pt/C catalysts for fuel cells.

REFERENCES

[1] Novoselov, K. S.; Geim, A. K.; Morozov, S. V.; Jiang, D.; Zhang, Y.; Dubonos, S. V.; Grigorieva, I. V.; Firsov, A. A., *Science* 2004, *306*, 666.
[2] Geim, A. K.; Novoselov, K. S., *Nat. Mater.* 2007, *6*, 183.
[3] Stoller, M. D.; Park, S. J.; Zhu, Y. W.; An, J. H.; Ruoff, R. S., *Nano Lett.* 2008, *8*, 3498.
[4] Lee, C.; Wei, X.; Kysar, J. W.; Hone, J., *Science* 2008, *321*, 385.

[5] Balandin, A. A.; Ghosh, S.; Bao, W. Z.; Calizo, I.; Teweldebrhan, D.; Miao, F.; Lau, C. N., *Nano Lett.* 2008, *8*, 902.
[6] Liu, M. M.; Liu, R.; Chen, W., *Biosens. Bioelectron.* 2013, *45*, 206.
[7] Liu, M. M.; Chen, W., *Biosens. Bioelectron.* 2013, *46*, 68.
[8] Kou, R.; Shao, Y. Y.; Mei, D. H.; Nie, Z. M.; Wang, D. H.; Wang, C. M.; Viswanathan, V. V.; Park, S.; Aksay, I. A.; Lin, Y. H.; Wang, Y.; Liu, J., *J. Am. Chem. Soc.* 2011, *133*, 2541.
[9] Cao, A. N.; Liu, Z.; Chu, S. S.; Wu, M. H.; Ye, Z. M.; Cai, Z. W.; Chang, Y. L.; Wang, S. F.; Gong, Q. H.; Liu, Y. F., *Adv. Mater.* 2010, *22*, 103.
[10] Zhang, S.; Shao, Y. Y.; Liao, H. G.; Liu, J.; Aksay, I. A.; Yin, G. P.; Lin, Y. H., *Chem. Mater.* 2011, *23*, 1079.
[11] Jin, Z.; Nackashi, D.; Lu, W.; Kittrell, C.; Tour, J. M., *Chem. Mater.* 2010, *22*, 5695.
[12] Liu, M. M.; Lu, Y. Z.; Chen, W., *Adv. Funct. Mater.* 2013, *23*, 1289.
[13] Lu, Y. Z.; Jiang, Y. Y.; Wu, H. B.; Chen, W., *J. Phys. Chem. C* 2013, *117*, 2926.
[14] Li, B.; Cao, X. H.; Ong, H. G.; Cheah, J. W.; Zhou, X. Z.; Yin, Z. Y.; Li, H.; Wang, J. L.; Boey, F.; Huang, W.; Zhang, H., *Adv. Mater.* 2010, *22*, 3058.
[15] Li, X. S.; Cai, W. W.; An, J. H.; Kim, S.; Nah, J.; Yang, D. X.; Piner, R.; Velamakanni, A.; Jung, I.; Tutuc, E.; Banerjee, S. K.; Colombo, L.; Ruoff, R. S., *Science* 2009, *324*, 1312.
[16] He, Q. Y.; Sudibya, H. G.; Yin, Z. Y.; Wu, S. X.; Li, H.; Boey, F.; Huang, W.; Chen, P.; Zhang, H., *Acs Nano* 2010, *4*, 3201.
[17] Guo, S. J.; Dong, S. J., *Chem. Soc. Rev.* 2011, *40*, 2644.
[18] Lotya, M.; Hernandez, Y.; King, P. J.; Smith, R. J.; Nicolosi, V.; Karlsson, L. S.; Blighe, F. M.; De, S.; Wang, Z. M.; McGovern, I. T.; Duesberg, G. S.; Coleman, J. N., *J. Am. Chem. Soc.* 2009, *131*, 3611.
[19] Chen, A. C.; Holt-Hindle, P., *Chem. Rev.* 2010, *110*, 3767.
[20] Chen, W.; Chen, S. W., *Angew. Chem. Int. Edit.* 2009, *48*, 4386.
[21] Chen, W.; Chen, S. W., *J. Mater. Chem.* 2011, *21*, 9169.
[22] Chen, W.; Kim, J. M.; Sun, S. H.; Chen, S. W., *Langmuir* 2007, *23*, 11303.
[23] Biddinger, E. J.; von Deak, D.; Ozkan, U. S., *Top. Catal.* 2009, *52*, 1566.
[24] Casanovas, J.; Ricart, J. M.; Rubio, J.; Illas, F.; JimenezMateos, J. M., *J. Am. Chem. Soc.* 1996, *118*, 8071.

[25] Kundu, S.; Nagaiah, T. C.; Xia, W.; Wang, Y. M.; Van Dommele, S.; Bitter, J. H.; Santa, M.; Grundmeier, G.; Bron, M.; Schuhmann, W.; Muhler, M., *J. Phys. Chem. C* 2009, *113*, 14302.

[26] Wu, M.; Cao, C.; Jiang, J. Z., *Nanotechnology* 2010, *21*.

[27] Deifallah, M.; McMillan, P. F.; Cora, F., *J. Phys. Chem. C* 2008, *112*, 5447.

[28] Wei, D. C.; Liu, Y. Q.; Wang, Y.; Zhang, H. L.; Huang, L. P.; Yu, G., *Nano Lett.* 2009, *9*, 1752.

[29] Qu, L. T.; Liu, Y.; Baek, J. B.; Dai, L. M., *Acs Nano* 2010, *4*, 1321.

[30] Jin, Z.; Yao, J.; Kittrell, C.; Tour, J. M., *Acs Nano* 2011, *5*, 4112.

[31] Zhang, C. H.; Fu, L.; Liu, N.; Liu, M. H.; Wang, Y. Y.; Liu, Z. F., *Adv. Mater.* 2011, *23*, 1020.

[32] Choucair, M.; Thordarson, P.; Stride, J. A., *Nat. Nanotechnol.* 2009, *4*, 30.

[33] Subrahmanyam, K. S.; Panchakarla, L. S.; Govindaraj, A.; Rao, C. N. R., *J. Phys. Chem. C* 2009, *113*, 4257.

[34] Sheng, Z. H.; Shao, L.; Chen, J. J.; Bao, W. J.; Wang, F. B.; Xia, X. H., *Acs Nano* 2011, *5*, 4350.

[35] Yang, S. B.; Feng, X. L.; Wang, X. C.; Mullen, K., *Angew. Chem. Int. Edit.* 2011, *50*, 5339.

[36] Yang, S. B.; Zhi, L. J.; Tang, K.; Feng, X. L.; Maicr, J.; Mullen, K., *Adv. Funct. Mater.* 2012, *22*, 3634.

[37] Lai, L. F.; Potts, J. R.; Zhan, D.; Wang, L.; Poh, C. K.; Tang, C. H.; Gong, H.; Shen, Z. X.; Jianyi, L. Y.; Ruoff, R. S., *Energ. Environ. Sci.* 2012, *5*, 7936.

[38] Zheng, Y.; Jiao, Y.; Ge, L.; Jaroniec, M.; Qiao, S. Z., *Angew. Chem. Int. Edit.* 2013, *52*, 3110.

[39] Endo, M.; Hayashi, T.; Hong, S. H.; Enoki, T.; Dresselhaus, M. S., *J. Appl. Phys.* 2001, *90*, 5670.

[40] Martins, T. B.; Miwa, R. H.; da Silva, A. J. R.; Fazzio, A., *Phys. Rev. Lett.* 2007, *98*.

[41] Lherbier, A.; Blase, X.; Niquet, Y. M.; Triozon, F.; Roche, S., *Phys. Rev. Lett.* 2008, *101*.

[42] Lin, T. Q.; Huang, F. Q.; Liang, J.; Wang, Y. X., *Energ. Environ. Sci.* 2011, *4*, 862.

[43] Li, X.; Fan, L. L.; Li, Z.; Wang, K. L.; Zhong, M. L.; Wei, J. Q.; Wu, D. H.; Zhu, H. W., *Adv. Energy Mater.* 2012, *2*, 425.

[44] Lu, X. J.; Wu, J. J.; Lin, T. Q.; Wan, D. Y.; Huang, F. Q.; Xie, X. M.; Jiang, M. H., *J. Mater. Chem.* 2011, *21*, 10685.

[45] Sheng, Z. H.; Gao, H. L.; Bao, W. J.; Wang, F. B.; Xia, X. H., *J. Mater. Chem.* 2012, *22*, 390.

[46] Wang, S. Y.; Zhang, L. P.; Xia, Z. H.; Roy, A.; Chang, D. W.; Baek, J. B.; Dai, L. M., *Angew. Chem. Int. Edit.* 2012, *51*, 4209.

[47] Yang, L. J.; Jiang, S. J.; Zhao, Y.; Zhu, L.; Chen, S.; Wang, X. Z.; Wu, Q.; Ma, J.; Ma, Y. W.; Hu, Z., *Angew. Chem. Int. Edit.* 2011, *50*, 7132.

[48] Zhang, C. Z.; Mahmood, N.; Yin, H.; Liu, F.; Hou, Y. L., *Adv. Mater.* 2013, *25*, 4932.

[49] Yang, Z.; Yao, Z.; Li, G. F.; Fang, G. Y.; Nie, H. G.; Liu, Z.; Zhou, X. M.; Chen, X.; Huang, S. M., *Acs Nano* 2012, *6*, 205.

[50] Yao, Z.; Nie, H. G.; Yang, Z.; Zhou, X. M.; Liu, Z.; Huang, S. M., *Chem. Commun.* 2012, *48*, 1027.

[51] Guo, S. J.; Sun, S. H., *J. Am. Chem. Soc.* 2012, *134*, 2492.

[52] Zhang, R. Z.; Chen, W., *J. Mater .Chem. A* 2013, *1*, 11457.

[53] Wang, H. L.; Liang, Y. Y.; Li, Y. G.; Dai, H. J., *Angew. Chem. Int. Edit.* 2011, *50*, 10969.

[54] Liang, Y. Y.; Li, Y. G.; Wang, H. L.; Zhou, J. G.; Wang, J.; Regier, T.; Dai, H. J., *Nat. Mater.* 2011, *10*, 780.

[55] Guo, S. J.; Zhang, S.; Wu, L. H.; Sun, S. H., *Angew. Chem. Int. Edit.* 2012, *51*, 11770.

[56] Kim, S. K.; Jeon, S., *Electrochem. Commun.* 2012, *22*, 141.

[57] Lefevre, M.; Dodelet, J. P.; Bertrand, P., *J. Phys. Chem. B* 2002, *106*, 8705.

[58] Baranton, S.; Coutanceau, C.; Roux, C.; Hahn, F.; Leger, J. M., *J. Electroanal. Chem.* 2005, *577*, 223.

[59] Dong, G. F.; Huang, M. H.; Guan, L. H., *Phys. Chem. Chem. Phys.* 2012, *14*, 2557.

[60] Jahan, M.; Bao, Q. L.; Loh, K. P., *J. Am. Chem. Soc.* 2012, *134*, 6707.

[61] Jiang, Y. Y.; Lu, Y. Z.; Lv, X. Y.; Han, D. X.; Zhang, Q. X.; Niu, L.; Chen, W., *Acs Catal.* 2013, *3*, 1263.

[62] Kamiya, K.; Hashimoto, K.; Nakanishi, S., *Chem. Commun.* 2012, *48*, 10213.

[63] Zhang, C. Z.; Hao, R.; Yin, H.; Liu, F.; Hou, Y. L., *Nanoscale* 2012, *4*, 7326.

[64] Tsai, C. W.; Tu, M. H.; Chen, C. J.; Hung, T. F.; Liu, R. S.; Liu, W. R.; Lo, M. Y.; Peng, Y. M.; Zhang, L.; Zhang, J. J.; Shy, D. S.; Xing, X. K., *Rsc Adv.* 2011, *1*, 1349.

[65] Li, S. Z.; Hu, Y. Y.; Xu, Q.; Sun, J.; Hou, B.; Zhang, Y. P., *J. Power Sources* 2012, *213*, 265.

[66] Parvez, K.; Yang, S. B.; Hernandez, Y.; Winter, A.; Turchanin, A.; Feng, X. L.; Mullen, K., *Acs Nano* 2012, *6*, 9541.
[67] Byon, H. R.; Suntivich, J.; Shao-Horn, Y., *Chem. Mater.* 2011, *23*, 3421.
[68] Wu, Z. S.; Yang, S. B.; Sun, Y.; Parvez, K.; Feng, X. L.; Mullen, K., *J. Am. Chem. Soc.* 2012, *134*, 9082.
[69] Kostecki, R.; Richardson, T.; McLarnon, D., *J. Electrochem. Soc.* 1998, *145*, 2380.
[70] Rahim, M. A. A.; Hameed, R. M. A.; Khalil, M. W., *J. Power Sources* 2004, *134*, 160.
[71] Zhang, L. R.; Zhao, J.; Li, M.; Ni, H. T.; Zhang, J. L.; Feng, X. M.; Ma, Y. W.; Fan, Q. L.; Wang, X. Z.; Hu, Z.; Huang, W., *New J. Chem.* 2012, *36*, 1108.
[72] Elzatahry, A. A.; Abdullah, A. M.; El-Din, T. A. S.; Al-Enizi, A. M.; Maarouf, A. A.; Galal, A.; Hassan, H. K.; El-Ads, E. H.; Al-Theyab, S. S.; Al-Ghamdi, A. A., *Int. J. Electrochem. Sc.* 2012, *7*, 3115.
[73] Wang, Z. H.; Du, Y. L.; Zhang, F. Y.; Zheng, Z. X.; Zhang, Y. Z.; Wang, C. M., *J. Solid State Electr.* 2013, *17*, 99.
[74] Roche, I.; Chainet, E.; Chatenet, M.; Vondrak, J., *J. Phys. Chem. C* 2007, *111*, 1434.
[75] Xiao, W.; Wang, D. L.; Lou, X. W., *J. Phys. Chem. C* 2010, *114*, 1694.
[76] Yang, Y.; Shi, M.; Li, Y. S.; Fu, Z. W., *J. Electrochem. Soc.* 2012, *159*, A1917.
[77] Lee, J. S.; Lee, T.; Song, H. K.; Cho, J.; Kim, B. S., *Energ. Environ. Sci.* 2011, *4*, 4148.
[78] Wang, H. L.; Yang, Y.; Liang, Y. Y.; Zheng, G. Y.; Li, Y. G.; Cui, Y.; Dai, H. J., *Energ. Environ. Sci.* 2012, *5*, 7931.
[79] Yan, X. Y.; Tong, X. L.; Zhang, Y. F.; Han, X. D.; Wang, Y. Y.; Jin, G. Q.; Qin, Y.; Guo, X. Y., *Chem. Commun.* 2012, *48*, 1892.

In: Graphene
Editor: Bruce T. Edwards
ISBN: 978-1-62948-795-3

Chapter 4

APPLICATION OF GRAPHENE IN MASS SPECTROMETRY

Xianglei Kong*

The State Key Laboratory of Elemento-organic Chemistry,
Nankai University, Tianjin, China

ABSTRACT

This chapter reviews the up-to-date research about the application of graphene and its relatives in mass spectrometry (MS). Due to its large specific surface area, large delocalized π-electrons, thermal conductivity, stability and rich interaction chemistry, graphene has been widely used in MS study. Graphene and graphene-based materials were applied as very effective matrices for many kinds small molecules in matrix-assisted laser desorption/ionization (MALDI) MS analysis. Many advantages of this novel matrix have been proved, which included: low interference ions from matrix itself, good reproducibility, high salt tolerance, less accompanied reduction and so on. The unique properties of graphene also make it a superior sorbent used in solid-phase extraction (SPE). Further development of on-line SPE methods based on graphene coupled with MALDI-MS, GC-MS and LC-MS greatly simplifies the analytical procedure for complex samples and makes the high-throughput and automatic analysis possible. And its potential in rapid analysis of environmental contaminants has been gradually revealed by more and

* Email: kongxianglei@nankai.edu.cn.

more research groups. Their applications as a platform in proteolysis for the rapid identification of proteins are also presented here. In addition, laser ablation mass spectra of graphene and graphene oxide have been studied. Graphene was also found to be a unique precursor for the generation of large-sized carbon cluster anions in the gas phase.

1. INTRODUCTION

Since its isolation and characterization by Novoselov et al. in 2004, graphene has been extensively studied in the last ten years by scientists in different research areas. [1-8] Graphene is a two-dimensional single layer of carbon atoms bound in a hexagonal lattice structure. Due to its unique structure, graphene shows many exceptional properties, such as large surface area, high values of Young's modulus, high thermal conductivity, excellent optical transparency and others. Because of this, graphene has been a promising material in its applications in many fields, including high-speed and radio frequency logic devices, sensors, displays, solar cells and catalysis [2-8].

Chemists in mass spectrometry (MS) related fields soon realized that this new material might be applied as a new tool in analytical chemistry. In fact, this is not the first time to think such a problem. After the macroscopic generation of fullerenes and carbon nanotubes (CNTs), they were successfully applied as sorbents for solid-phase extraction (SPE) or solid-phase microextraction (SPME). [9-12] These carbon nanomaterials were also used as matrices for the analysis of some small molecules by the method of matrix-assisted laser desorption/ionization (MALDI) MS. [13-26] Due to its very specific surface area, it is natural to suggest that graphene should be a better SPE material than fullerene and CNT. And many experimental results obtained recently do support this point. [27] In addition, the large delocalized π-electron system of graphene provides a strong affinity for carbon-based ring structures, which indicates that graphene may play a very unique and important role in the analysis of relative pollutants or drugs from complex environmental and biological samples.

In this chapter, we will highlight the recent applications of graphene and its derivatives in MS-related chemistry. The topics include their application in the laser desorption/ionization (LDI) MS, in the sample preparation methods for rapid MS analysis and in the study of carbon cluster ions. The general methods for the synthesis and characterization of graphene and its derivatives will not be discussed here. However, in some cases, readers need to read the corresponding references to know more details about it, since it is well known that the synthesis methods of graphene might have some important influence on its performance.

2. Application of Graphene and Its Relatives in LDI-MS

The method of LDI was introduced for MS analysis in the late 1960s, and it has been a useful tool for the rapid analysis of many kinds of light-absorbing organic compounds. By mixing some strongly light-absorbing organic molecules to the sample prior to the LDI analysis can greatly enhance the corresponding detect sensitivity and extend its application in many aspects. The method is known as MALDI (Figure 1). [28,29] By now, many kinds of matrices have been developed and applied. Due to the high sensitivity and great versatility of the method, MALDI MS has been a major analytical tool in modern sciences. However, most of these applications are aimed at biomolecules with mass larger than 500 u. The direct analysis of small molecules by MALDI MS is still difficult due to the matrix interference in the low-mass region. [30] In order to solve this problem, kinds of new matrices including inorganic nanoparticles and some designed organic molecules have been developed and applied. [30] And an alternative method, known as surface-enhanced laser desorption/ionization (SELDI), has also been developed to solve these problems (Figure 1c). [31].

During the past three or four years, graphene and its relatives have been widely used as a new matrix or surface to be applied in MALDI and SELDI MS (Table I). [32-54] Graphene was firstly used as a good detection platform for MS analysis of single-stranded DNA molecules by Tang et al.. [32] It has been found that graphene could produce lower fragmentation products in the SELDI MS than graphene oxide (GO). With the help of graphene, very low limits of detection (LODs) were achieved for DNA sample (100 fM) and protein sample (1pM). Dong X. et al. have found that graphene was a good matrix for the analysis of low molecular weight compounds using MALDI-MS.[33] Polar compounds including amino acids, nucleosides, and nonpolar compounds such as steroids, have been successfully detected with graphene matrix. Sodiated ions of analytes, instead of protonated ions, were observed in the positive ion mode with high resolution.

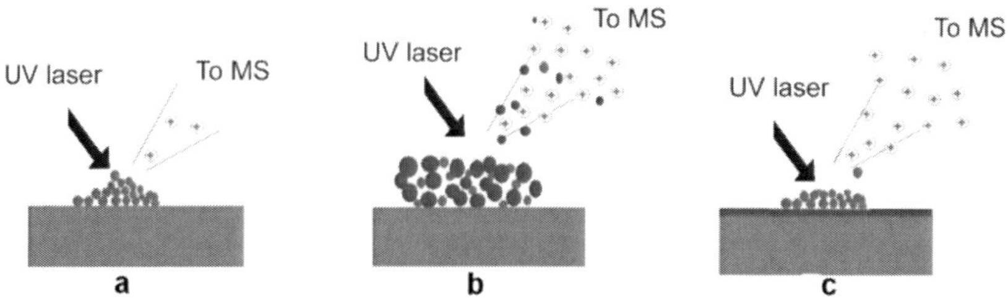

Figure 1. Schematic drawings that illustrate different laser desorption/ionization (LDI) sources for mass spectroscopic analysis: (a) direct LDI, (b) matrix assisted LDI (MALDI) and (c) surface enhanced LDI (SELDI).

Table I. Graphene and it relatives applied in MALDI MS as a matrix or SELDI MS as a surface

Materials applied	Analyte	Ion mode	Method	Year	Ref
Graphene/ GO	ssDNA, Cyt C, benzylpyridinium	positive	SELDI/MALDI	2010	[32]
Graphene	amino acids, polyamines, anticancer drugs, nucleosides	positive	MALDI	2010	[33]
RGO films	OCDD	negative	MALDI	2010	[34]
Graphene	amino acids, peptides, fatty acids, nucleosides	negative	MALDI	2011	[35]
Graphene	polymers	positive	SELDI	2011	[36]
Graphene /GO	components from traditional Chinese medicine herbs	positive/ negative	MALDI	2011	[37]
Graphene /GO	long-chain fatty acids	positive	MALDI	2011	[38]
Graphene /GO	tetracyclines	positive	MALDI	2012	[39]
Graphene	polycyclic aromatic hydrocarbons (PAH)	positive	MALDI	2011	[40]
Graphene	redox-sensitive molecules	positive	MALDI	2013	[41]
Mildly oxidized graphene	environmental pollutants	negative	MALDI	2013	[42]
GO	flavonoids, derivatives of coumarin	negative	SELDI	2012	[43]
GO	acetylcholine	positive	MALDI	2011	[44]

Materials applied	Analyte	Ion mode	Method	Year	Ref
GO	peptides	positive	MALDI	2012	[45]
Graphene	tripeptide glutathione	positive	MALDI	2013	[46]
Graphene	metal-drug complexes	positive	MALDI	2012	[47]
GO/carbon nanotube films	phospholipase	positive	SELDI	2010	[48]
GO/MWCNT films	small molecules and tissue imaging	positive/ negative	SELDI	2011	[49]
Multilayer films of GO and carbon nanotube	small molecules including glucose, lysine, cellobiose, etc.	positive	SELDI	2012	[50]
Au-GO hybrid film	small molecules including glucose, mannitol, cellobiose, etc.	positive	SELDI	2012	[51]
Graphene-coated cobalt nanoparticles	pentachlorophenol, bispenol A, polyfluorinated compounds, etc	positive/ negative	SELDI	2012	[52]
Magnetic graphene	traditional Chinese medicine, nicotine nitrogen oxides	positive	MALDI	2012	[53]
Aptamer-conjugated GO	cocaine, adenosine	positive	SELDI	2010	[54]
Laser Engineered Graphene Paper	a variety of small molecules	positive	SELDI	2013	[55]

Other samples including tetracyclines and polycyclic aromatic hydrocarbons (PAHs) have been also detected with the matrices of graphene or GO. [37-48] Lu et al. have found that graphene can be used for characterization of various polymers including polypropylene glycol and poly styrene. [36] It has been found that graphene matrix could not only eliminate the interference of matrix background ions, avoid the fragmentation of analytes and thus simplify the mass spectrum, but also improve shot-to-shot reproducibility. And its high salt tolerance revealed its potential usage in rapid analysis of small molecules in real biosamples.

Differently, graphene was also found to be a very effective matrix in the negative ion mode. Based on this advantage, Lu et al. used graphene flakes to fulfill the MS analysis of a wide range of small molecules including peptides and fatty acids. [35] Reduced graphene oxide (RGO) film has been applied as the matrix for the detection of octachlorodibenzo-p-dioxin (OCDD) by Zhou X. et al.. [34] Ions of $[M\text{-}Cl]^-$ of OCDD were detected in the negative ion mode, and the detection limit was as low as 500 pg.

It is found that the graphene matrix can minimize reduction in the process of MALDI for some redox-sensitive molecules such as riboflavin and methylene blue. [41] Although MALDI has been widely applied as a method of soft ionization, it has also been found that reduction is an inherent feature accompanying this process when organic matrices such as α-cyano-4-hydroxycinnamic acid (CHCA) and 2,5-dihydroxybenzoic acid (DHB) are used. [41] Due to this, it is thought that MALDI results cannot be directly used to judge the reduction states of related biosamples. For graphene matrix, the reduction can be minimized due to the lack of hydrogen radicals or protons in the plume formed by laser irradiation. Figure 2 shows the mass spectra of riboflavin obtained with the matrix of CHCA and graphene, respectively. Instead of protonated ions, the sodiated and potassiated ions of riboflavin are clearly detected with the graphene matrix. Remarkably, none reduced-species of $[MH+Na]^+$ or $[MH+K]^+$ was found in the mass spectrum.

As we mentioned above, many alternative inorganic materials, including graphite, fullerene and CNTs, have been tried and used as matrices in MALDI or SELDI MS analysis before the application of graphene matrix. Thus, it is interesting to see if the hybrid materials based on graphene and other nanoparticles can be used to as a matrix in MALDI or a surface in SELDI to get better performance. Kim et al. reported the fabrication of supported double-layer films of RGO and CNT on a transparent electrode. [56,57] These GO/CNT double-layer films were employed as a good platform for the SELDI MS analysis of phospholipids. [49] Time-dependent hydrolysis reactions of dipalmitoyphosphatidylcholine (DPPC) catalyzed by two different types of phospholipase: phospholipase A2 (PLA2) and phospholipase C (PLC) were tested with this platform. The enzyme solution was mixed with DPPC in a reaction vessel. Then 1μL aliquots were taken out and spotted onto the GO/CNT chip and analyzed by LDI-MS at intervals of 10 min. According to their relative intensity recorded in the mass spectra, the relative amount of each component was calculated and plotted against the reaction time. The versatile method can be used not only in the time-dependent assays, but also in the enzyme-concentration-dependent and inhibition assays.

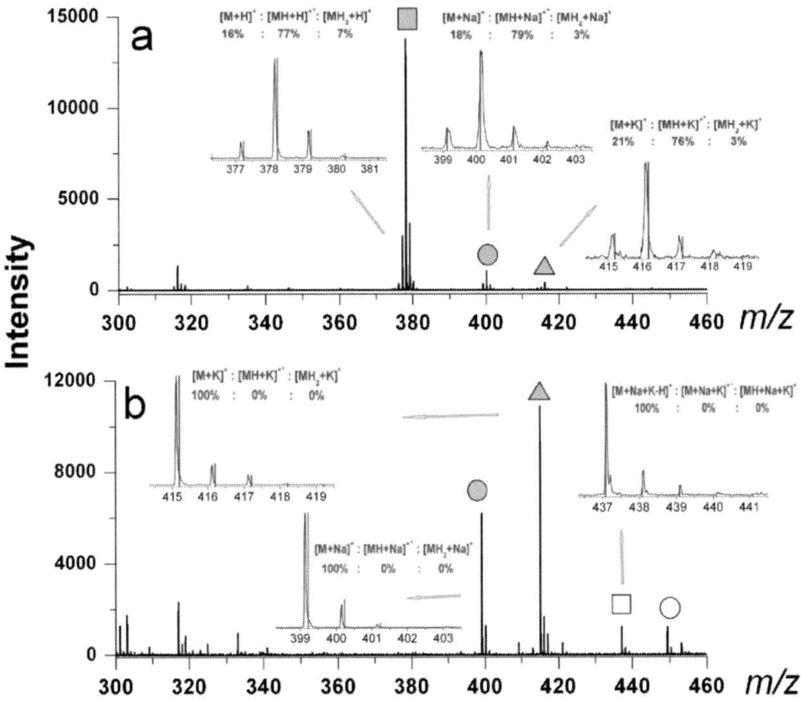

Figure 2. MALDI mass spectra of riboflavin obtained using matrices of (a) CHCA, (b) graphene. The protonated, sodiated and potassiated ions of riboflavin are marked by solid squares, circles and triangles. Other ions of $[M+Na+K-H]^+$ and $[M+2K-H]^+$ are identified by blank squares and circles, respectively, in (b). The isotropic distribution of each peak is shown in its corresponding inset. The bars are the simulated results, in which the ratios of different species used are shown in the upper right corner of each inset.

A surface-immobilized double-layer format through the self-assembly of GO sheets and aminated multiwalled (MW) CNT on a chemically modified substrate was also developed by the same group. [50] Its application in MALDI imaging MS of lipids distributed in mouse brain tissue was performed. The detecting efficiency of small molecules was affected by several aspects of the film, including thickness, surface roughness and assembly sequence. By controlling the processes of assembly cycles, it is possible to fulfill the effective analysis of target molecules based on MS or imaging MS under optimized conditions.

Metal nanoparticles have been applied to form hybrid films with graphene, too. Kim et al. prepared gold/GO hybrid films by directly growing gold nanoparticles on the supported thin layers of GO films on a glass slide. [52] With this hybrid film, various small molecules were analyzed using the method of SELDI MS. The ionization efficiency was greatly enhanced by the dense gold nanostructures and the underling GO films dissipated excess thermal energy by laser irradiation of Au rapidly to prevent fragmentation of analyte. Graphene-coated cobalt nanoparticles were prepared via flame spray pyrolysis by Kawasaki et al.. [53] The surface of these nanomagnets was further functionalized with benzylamine groups. Various small molecule drugs and environmentally significant compounds were successfully analyzed at high sensitivities with the help of these modified nanomagnets.

A new type of graphene paper was prepared by Qian et al. and the stability of it could be dramatically increased through a pulsed-laser engineering process. [55] Though the process, both surface hydrophobicity and electric conductivity of the paper have been enhanced or improved. This engineered graphene paper was applied as a surface of SELDI-MS detection and MS imaging. Compared to the pristine graphene paper or other commercial products, the engineered graphene paper increases the detection limits of a variety of small molecules by over two orders, which shows its great potential in diverse MS-based applications such as diagnostics, biomarker discovery and forensic examination.

Briefly, compared with conventional matrices, graphene shows some unique advantages in its application of MALDI and SELDI MS. First, it can be directly applied in small molecular analysis by MALDI and SELDI MS due to its advantage of interference-free in the low-molecular mass region. Second, its advantage of reduced spot-to-spot difference or good reproducibility, makes its potential in quantitative analysis based on MS. Third, the high salt tolerance makes it suitable in direct analysis of complicated biological and environmental samples. Fourth, its great performance in negative ion mode, makes its application in some environmental pollutants can be performed readily. Fifth, the minimization of accompanied reduction in normal MALDI process makes the detection of some redox-sensitive molecules more accurate. Sixth, the fabricated hybrid platform of graphene and other materials is very stable and durable against mechanical and photoagitation. Its reusability makes it more cost-effective than other existing SELDI platforms.

3. Application of Graphene and Its Relatives in SPE-MS

3.1. SPE-MALDI-MS

Graphene holds great promise as an adsorbent in SPE or SPME due to its large surface area, delocalized π-electrons and performability of surface modification. This point causes that graphene can play a dual role in the SPE-MALDI-MS. In this kind of application, graphene sheets or its relatives are used to enrich analytes selectively from complicated samples first, and then they were separated from the mixtures and subject to MALDI-MS analysis directly. Usually, the elution step can be omitted since the graphene was used as the matrix in the SPE–MALDI-MS. This point makes the whole analytical process more rapid and convenient. Surely an effective separation of the graphene from the diluted solution or complex biosamples after the absorption and enrichment of the analyte is very important. Typically, the pellets of graphene can be collected by centrifugation at 13 000 rpm for 5-10 min and further removing of the supernatant in the centrifuge tube (Figure 3). [32,33] After the separation, the graphene needs to be washed once or more by water or other solvents before MALDI MS analysis. [46, 58] In order to minimize the loss of graphene in such a progress and accelerate the progress, magnetic graphene were made by combing Fe_3O_4 or cobalt particles with the graphene sheets. [46,53,54] Thus, these analyte-bound magnetic graphene particles can be readily separated from the supernatant using an external magnetic field (Figure 3).

Experiments based on the SPE-MALDI-MS with graphene and its relatives as both adsorbent and matrix for various analytes are listed in Table II. The first example was shown by Tang et al.. They applied graphene as both a SPE material and a SELDI probe for DNA molecules. [32] Liu et al. have successfully applied graphene and GO to enrich long-chain fatty acids. [38] Graphene was also used to extract coronene from river water sample and the MALDI detection limit was 10^{-7} M. [40] Liu et al. applied graphene and GO to enrich tetracylines from milk samples, and the LOD in corresponding MALDI MS experiments could be as low as 2 nM. [39] Shi et al. fulfilled the enrichment of several small molecules including curcumin, luteoloside and nicotine nitrogen oxides from diluted water solutions using the magnetic graphene. [53] In their experiment, magnetic graphene served as both an adsorbent and matrix.

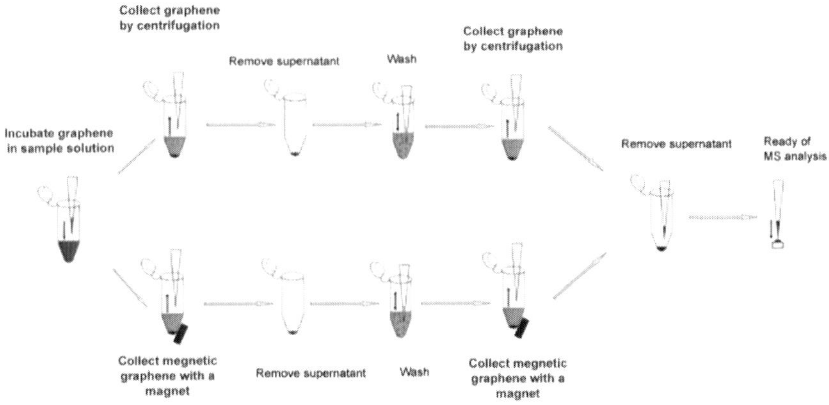

Figure 3. The application of graphene and magnetic graphene particles in SPE/SPME based MALDI MS analysis.

Table II. Experiments performed by the method of SPE-MALDI-MS with graphene and its relatives as adsorbents

Materials applied	Analyte	Elution	Separation	Year	Ref
Graphene	aqueous solution of ssDNA and protein	no	centrifugation	2010	[32]
Graphene	squalene from ethanol	no	centrifugation	2010	[33]
Graphene /GO	long-chain fatty acids from serum and urine samples	no	centrifugation	2011	[38]
Graphene /GO	tetracyclines from milk samples	no	centrifugation	2012	[39]
Graphene	PAHs from water	no	centrifugation	2011	[40]
Aptamer-conjugated GO	cocaine and adenosine from plasma samples	no	centrifugation	2010	[54]
Graphene - coated cobalt nanoparticles	aqueous solution of perfluorooctanesulfonate, pentachlorophenol, bisphenol A, and polyfluorinated compounds with varying chain length	no	magnetic separation	2012	[52]
Magnetic graphene	curcumin, luteoloside, nicotine nitrogen oxides from water	no	magnetic separation	2012	[53]
G@silica, GO@silica	chlorophenols, peptides, proteins from water	yes	SPE cartridge	2011	[58]
LaGM probe grafted with Fe_3O_4	peptides and phosphopeptides from serum and urine	yes	magnetic separation	2012	[45]

The whole experimental process was thus simplified. Graphene-coated cobalt nanoparticles were also used to extract perfluorooctanesulfonate from large volumes of aqueous solutions by magnetic separation by Kawasaki et al., and the corresponding detect sensitivity could go down to 0.1 ng/L [52].

To improve the selectivity of extraction, chemical modification of graphene is necessary. Gulbakan et al. prepared aptamer-conjugated GO and applied it for selective enrichment of cocaine and adenosine from plasma samples. [54] In the experiments, the thiol-functionalized cocaine and adenosine aptamers were conjugated to GO by activating the carboxyl groups. Poly(ethylene glycol) molecules were applied as the spacer unit. In a contrastive experiment, GO with physically adsorbed cocaine aptamer was used for target extraction. However, it has been found that the follow washing steps caused the signal of cocaine in MS totally decreased, indicating that the chemical conjugation of the aptamer is very necessary for the capture.

3.2. SPE-GC-MS and SPE-LC-MS

Graphene-based SPE and SPME methods are also combined with GC-MS and LC-MS for the analysis of many environmental molecules, which are summarized in Table III. [59-69] Chen et al. prepared a novel graphene based SPME fiber by immobilizing the graphene on a stainless steel wire. [70] The fiber was characterized by its homogeneous, porous and wrinkled surface and excellent thermal and chemical stability. Although the thickness of the coating is only 6-8 μm, its extraction efficiency for six kinds of pyrethroid pesticides in water samples was higher than those of commercial fibers. And its good mechanical stability also allows a long lifetime up to more than 250 extractions. A new SPME device for the GC-MS determination of polybrominated diphenyl ethers (PBDEs) was developed by Zhang et al.. [58] It was assembled in a plunger-in-needle microsyringe, of which the plunger was coated by graphene via a sol-gel approach. The excitation efficiency was greatly enhanced due to the porous, three-dimensional network structure with a large surface area created by the coating process. After the SPME process, the fiber was retracted into the needle and inserted into the GC injector at 300 ^{o}C for analyte desorption for 5 min. With GC-MS analysis, the procedure can achieve low detection limits ranging from 0.2 and 5.3 ng/L. The method offers a simple, rapid, sensitive and inexpensive tool for the determination of trace PBDEs in water samples.

Table III. Experiments performed by SPE-GC-MS and SPE-LC-MS

Materials applied	Analyte	LOD	Recovery	Elution	Detection	Year	Ref
Graphene-based sol–gel	polybrominated diphenyl ethers (PBDEs)	0.2-2.3 ng/L	75-82%	no	GC-MS	2011	[59]
Graphene-coated fiber	8 PAHs	1.52-2.72 ng/L	84-102% for water samples 72-95% for soil samples	no	GC-MS	2011	[60]
Graphene/polymer composite	16 PAHs	0.005-0.429 ng/mL	55-144%	no	GC-MS	2011	[61]
Graphene –supported zinc oxide fiber	sulfur volatiles	100-700 ng/L	73-81%	no	GC-MS	2012	[62]
Colloidal graphene	15 phthalate acid esters	2-7 μg/L	71-111%	yes	GC-MS	2013	[63]
Graphene	9 PAHs	0.84-13 ng/L	73-106%	yes	GC-MS	2013	[64]
Graphene	10 OH-PBDEs	5.3-212.6 pg/g	24-81% for fish samples 71-118% for soil samples 52-114% for tree bark samples	yes	LC-ESI-MS/MS	2011	[65]
Superparamagnetic nanoparticle-decorated graphene sheets	Perfluoroalkyl substances alkyltrimethylammonium	0.15-0.50 ng/L 1.4-8.0 ng/L	56-94%	yes	HPLC–ESI-MS/MS	2012	[66]
Monolithic column incorporated with graphene	9 glucocorticoids	0.13-1.93 ng/mL	84-104%	yes	LC-MS	2012	[67]
Graphene cartridge	chloramphenicol	0.036 μg/Kg	92-103%	yes	HPLC-MS/MS	2012	[68]
GO/G@poly(GMA-EDMA)	sarcosine	1.0 ng /mL	93%	yes	LC-MS/MS	2013	[69]

Based on a layer-by-layer fabrication method, Zhang et al. prepared a different SPME fiber coated with graphene. [60] With the treatment of NaOH and 3-aminopropyltriethoxysilane (APTES), amino groups were introduced to the surface of silica fiber. Then GO was covalently linked to the fiber via the reaction of the amino groups with the carboxyl groups on GO. By repeating the treatment of APTES and GO four times, a homogeneous coating with a thickness of 20 μm could be prepared. The chemical bonding between GO and the silica greatly improve its chemical stability. It showed good precision, low detection limit and wide linearity for the GC-MS analysis of PAHs. The same group also prepared a graphene-supported ZnO SPME fiber via the sol-gel approach for the analysis of sulfur volatiles. [62] The fiber showed high extraction selectivity and sensitivity towards sulfur compounds.

Luo et al. prepared a new poly(ethylene glycol dimethacrylate)/graphene composite by *in situ* polymerization. [61] Based on the combination of stir rod sorptive extraction and GC/MS, the determination of 16 PAHs in water samples was fulfilled. Wu et al. designed a simple and inexpensive method of dispersive SPE based on graphene. [63] Combined with GC-MS, 15 phthalate acid esters were analyzed and overall average recoveries were between 71 and 117%. Wang et al. made a graphene-based SPE disk and applied it in the separation and preconcentration of trace PAHs from environmental water samples. [64] The prepared disks exhibited high loading rate and anticlogging. In the experiments, the analytes extracted by graphene were eluted by cyclohexane and determined by GC-MS. High flow rate (30 mL/min) and sensitivity (0.84-13 ng/L) was achieved for the selected 9 PAHs.

Unlike the methods of SPE-MALDI-MS and SPE-GC-MS, the analyte must be eluted from graphene before the further analysis by LC-MS. Thus, the elution conditions in each experiment should be optimized carefully. Different analytical protocols using graphene-based SPE/SPME coupled with LC-MS analysis have been developed for the determination of kinds of analytes (Table III) [65-69].

Liu et al. developed a novel graphene-assisted matrix solid-phase dispersion method for extraction of hydroxylated (OH-) PBDEs from environmental samples. [65] It has been found that grinding the solid sample with chemically converted graphene power could yield a tight contact and sufficient dispersion of the sample matrix. In a two-step strategy, PBDEs and methyoxylated PBDEs were firstly eluted and determined by GC; and then OH-PBDEs were eluted by acetone and analyzed by LC-MS/MS. This method shows its advantages of better recoveries (compared with other sorbents and extraction techniques), reduced consumption of sorbent and solvent, short extraction time and good selectivity. The same group also synthesized superparamagnetic nanoparticle-decorated graphene (MG) sheets and used them as the support for hemimicelles/admicelles for SPE of different compounds. [66] Depending on solution pH, the MG sheets could be positively or negatively charged, thus allowed different modes of extraction to be performed.

First graphene-packed SPE cartridge was also made by the same group. [71] Eight chlorophenols (CPs) were extracted with this cartridge, then eluted and determined by HPLC with multi-wavelength UV detection. Wu et al. developed a protocol using their graphene cartridge coupled with HPLC-MS/MS to determinate chloramphenicol (CAP) in aquatic products. [68] The cartridge has also been applied to other analytes including glutathione from human plasma, neurotransmitters from rat brain, trace amounts of lead and chromium (III). [72-74] Tong et al. prepared a polymer monolith microextraction (PMME) material of GO/GN@poly(GMA-EDMA) by coating GO or graphene nanosheets onto the poly(glycidyl methacrylate-ethylene dimethacrylate) [69]. Coupled with LC-MS/MS, the method was successfully utilized to the analysis of a potential prostate cancer biomarker: sarcosine. The LOD of 1.0 ng/mL and the recovery of 93% have been achieved. Results show that the performance of graphene was superior to that of other adsorbents, such as C18 silica, graphitic carbon and CNTs. Its advantages of high sensitivity, good reproducibility, reusability, compatibility with various organic solvents have been proven.

4. Application of Graphene in the MS Analysis of Proteins and Peptides

For the analysis of proteins or peptides, graphene can be used as a good SPE material too, but surface modification of graphene is necessary in order to improve its selectivity. Thus after enrichment and separation, these protein or peptide-bound particles or sheets can be further studied by MALDI MS with other conventional organic matrices such as CHCA or DHB. [45,58] In most of these cases, the elution step can be omitted. For example, Cheng et al. applied the $LaPO_4$-GO composites to enrich the low–abundance peptides and phosphopeptides from human urine and serum. [45] If a graphene-based SPE cartridge is used, then surely the elution step is necessary. Liu et al. used G@silica- and GO@silica- packed cartridges to enrich proteins and peptides, and then the eluted proteins and peptides were analyzed by MALDI-MS with CHCA matrix readily [58].

Effective protein digestion is the key procedure prior to protein identification based on MS. However, a typical time for tryptic digestion usually needs 8 hours or more, which makes it is incompatible with high-throughput MS-based protein analysis. On the other hand, for target proteins in very low concentrations, the autolysis of protease interferes with protein identification very much. In order to solve the problem, on-plate proteolysis approaches were developed based on the immobilization of proteases on a solid surface. GO provides a very good choice for this goal due to its ultra-high surface area, stability and abundant functional groups on the surface. Bao et al. reported the immobilization of trypsin in the layer-by-layer coating of GO and chitosan on a piece of glass fiber. [75] The mild environmental conditions minimized the denaturation and autolysis of trypsin effectively. Several proteins were tested and the digestion time could be reduced to less than 10s. The obtained digests were then identified by MALDI-MS. Results show that the performance of this bioreactor was comparable to that of 12h in-solution tryptic digestion.

To maintain the activity of the enzyme, Xu et al. immobilized trypsin on GO sheets using polymers as molecular spacers. [76] Based on the trypsin-linked GO sheets, a novel microwave-assisted on-plate rapid digestion method (the digestion can be fulfilled in 15s) has been developed with subsequent analysis by MALDI-MS. Due to its high digestion efficiency and miniaturizing sample preparation, this approach offers great promise for rapid and high-throughput protein identification. Besides this, its minimal sample consumption also makes the method could be applicable to the identification of low-abundance proteins. Alternatively, Jiao et al. reported a novel strategy for efficient *in situ* protein digestion with a GO immobilized enzyme reactor. [77] Compared to the traditional digestion protocols, this strategy also shows some advantages, including lower protein concentration, higher sequence coverage and shorter processing time (a typical on-plate digestion experiment can be performed in 10 minutes). And this in situ digestion method also can be applied into MALDI tissue imaging MS.

5. Laser Ablation MS of Graphene

Laser ablation relates a process of removing material from a solid surface by irradiating it with a strong laser beam. Combined with MS, the method can provide many valuable information about not only the interactions of laser radiation with various materials, but also some novel ions and species. It is one of the most convenient experimental methods to find new clusters with magic numbers. Among them, carbon clusters are particularly attractive due to their fascinating structures, chemical properties and significant role in deposition and combustion processes. The example of the C_{60} does reflect how laser ablation MS can help people to find new chemistry. [78] By now, different carbon materials have been studied by this method. For example, Campbell et al. have studied the carbon clusters produced by UV laser ablation of polyimide using time-of-flight MS. [79] With a Fourier transform ion cyclotron resonance (FT ICR) MS, Greenwood et al. have observed the formation of carbon cluster ions by laser ablation of coals and other materials. [80] Sedo et al. have suggested that different carbon materials including graphite, diamond and carbon nanotubes are subjected to similar physic-chemical changes by UV-laser pulses. [81].

Using a FT ICR MS equipped with a MALDI source, Kong et al. have studied the laser ablation mass spectra of graphene and GO (Figure 4). [82] In the negative ion mode, they found that large even-numbered carbon cluster anions, C_n^-, with n up to 500, were generated. [95] It should be noticed that although positively charged carbon cluster ions in different sizes can be generated by laser ablation of various materials, the generation of carbon cluster anions of large sizes with n > 60 is very difficult and has been rarely reported. The signal intensity of the cluster anions generated from GO is much weaker than that generated from graphene. The cluster distributions generated from graphene can be altered by changing experimental conditions, such as laser energy and accumulation period in the FT ICR cell. Some weak signals of odd-numbered anions from C_{125}^- to C_{211}^- were also observed in the mass spectra. Tandem MS was applied to some selected cluster anions in the FT ICR cell. Very interestingly, it has been found that several cluster anions could lose units of C_2, C_4, C_6 or C_8 respectively in their collisionally activated dissociation (CAD) MS, but others could not.

The experiment was also performed in positive ion mode on the same instrument under similar experimental conditions. [83] For graphene, only narrow distributions of even-numbered clusters from C_{60}^+ to C_{162}^+ were observed, while for GO and other carbon materials like nanodiamonds, broad distributions of even-numbered clusters from C_{100}^+ to C_{400}^+ and weak signals of odd-numbered clusters from C_{189}^+ to C_{395}^+ were observed. The threshold of laser energy for graphene is found to belower than that of GO and nanodiamonds. And the effects of laser energy on distributions of product ions are also different for graphene and GO.

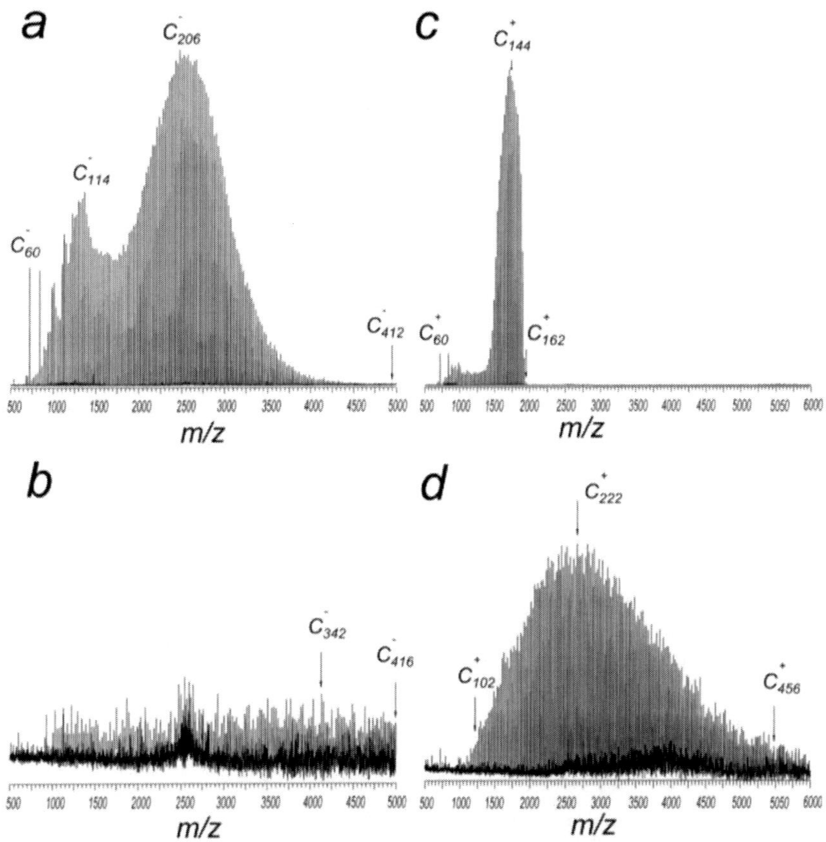

Figure 4. Laser ablation mass spectra of (a) graphene and (b) GO in the negative ion mode, and those of (c) graphene and (d) GO in the positive ion mode.

These results indicate the mechanisms for the formation of cluster ions in the laser ablation of graphene and other carbon materials are quite different. A top-down mechanism including both direct transformation of graphene to fullerene verified by Chuvilin et al. previously based on their experimental images and fragmentation of large-sized fullerenes is suggested for the process of laser ablation of graphene. [83] However, the results of GO are close to those of NDs, thus can be explained by the generally accepted bottom-up mechanism, in which the fullerene cages are considered to be formed from carbon atoms or small clusters [85-88].

CONCLUSION

They unique and fascinating properties of graphene, including lager surface area, good thermal and chemical stability, strong UV absorption and π-electron- rich structure, make it one of the most promising materials in the mass spectrometry. Graphene and graphene derivatives were used as a novel matrix or surface for MALDI-MS or SELDI-MS analysis of low-molecular-weight molecules. Compared with conventional matrices, graphene and its derivatives show many advantages, such as: reduced matrix-interference in low mass region, homogeneous distribution and good reproducibility, high salt tolerance, minimization of the accompanied reduction and so on. Graphene also shows its superior qualities as a sorbent for a variety of molecules from small PAHs to proteins in methods of SPE and SPME. Thus, graphene can play both roles of adsorbent and matrix for analyte enrichment and ionization in the combined method of SPE-MALDI-MS. This protocol not only simplifies analytical processes, but also shortens the time consumed and makes the identification of targets more effective and rapid. The graphene-coated fiber also makes the analysis based on SPE-GC-MS more sensitive and accurate. If an elution step is performed for the analyte-bonded graphene, other analytical methods, such as LC-MS, can be applied directly.

In spite of these successes in its application in mass spectrometry, many challenges remain and need to be solved or improved very much for graphene. First, current methods for producing graphene are still laborious, expensive and time-consuming. The success of graphene in new applications will surely promote the development of more convenient, low-cost and environmentally-friendly methods for the synthesis of high-quality graphene. [89-96] It is also very attractive to design new ways to make the product graphene more homogeneous in size and shape. Second, the selectivity of graphene applied in the SPE/SPME method should be improved. To get a better enrichment for the target analytes in very low concentration from solutions with large volume, the non-specific adsorption should be restricted further. To achieve this, more effective surface functionalization, material hybridization, or precise control in the size and shape need to be studied further. Third, there are still a lot of chances about the study of integration of graphene with different on-line MS methods. Although there are some great progresses in the graphene-based MS imaging and on-plate proteolysis, these techniques still need further improvement and more application cases to show its practicability and superiority in analytical chemistry. Fourth, application of graphene in the other MS-related researching field, such as cluster chemistry, gas phase chemistry and interface chemistry should be studied further, since a better understanding of the interactions between graphene and other molecules or laser irradiation should be very helpful for designing new analytical devices or methods and discovering their brand new applications in different researching fields.

REFERENCES

[1] K. S. Novoselov, A. K. Geim, S. V. Morozov, D. Jiang, Y. Zhang, S. V. Dubonos, I.V. Grigorieva, A. S. Firsov, *Science*. 306, 666 (2004).
[2] M. J. Allen, V. C. Tung, R. B. Kaner, *Chem. Rev.* 110, 132 (2010).
[3] S. J. Guo, S. J. Dong, *Chem. Soc. Rev.* 40, 2644 (2011).
[4] X. An, J. C. Yu, *RSC Advances,* 1, 1426 (2011).
[5] Y. wang, Y. Huang, Y. Song, X. Zhang, Y. Ma, J. Liang,Y. Chen, *Nano Lett.* 9, 220 (2009).
[6] D. Sui, Y. Huang, J. Liang, L. Huang, Y. Ma, Y. Chen, *Small* 7, 3186 (2011).
[7] X. Wan, Y. Huang, Y. Chen, *Accounts Chem. Res.* 45, 598 (2012).
[8] L. Huang, N. Yi, Y. Wu, Y. Zhang, Q. Zhang, Y. Huang, Y. Ma, Y. Chen, *Adv. Mater.* 25, 2224 (2013).
[9] M. Gallego, Y. P. Depena, M. Valcarcel, *Anal. Chem.* 66, 4074 (1994).
[10] J. R. Baena, M. Gallego, M. Valcarcel, *Trends Anal. Chem.* 21, 187 (2002).
[11] Y. Q. Cai, G. B. Jiang, J. F. Liu, Q. X. Zhou, *Anal. Chem.* 75, 2517 (2003).
[12] M. Valcarcel, S. Cardenas, B. M. Simonet, Y. Moliner-Martinez, R. Lucena, *Trends Anal. Chem.* 27, 34 (2008).
[13] L. Michalak, K. Fisher, D. Alderdice, D. Jardine, G. Willet, *Org. Mass Spectrom.* 29,512 (1994).
[14] F. Hopwood, L. Michalak, D. Alderdice, K. Fisher, G.Willet, *Rapid Commun. Mass Spectrom.* 8, 881(1994).
[15] X. Su, Y. Li, H. Zou, J. Qiu, Z. Guo, B. Guo, *Anal. Chem.* 75, 6191 (2003).
[16] J. T. Shiea, J. P. Huang, C. F. Teng, J. Y. Jeng, L. Y. Wang, L. Y. Chiang, *Anal. Chem.* 75, 3587 (2003).

[17] W. Chen, L. Wang, H. Chiu, Y. Chen, *J. Am. Soc. Mass Spectrom.* 15, 1629(2004).
[18] C. Pan, S. Xu, L. Hu, X. Su, J. Ou, H. Zou, Z. Guo, Y. Zhang, B. Guo, *J. Am. Soc. Mass Spectro.* 16, 883 (2005).
[19] M. Ugarov, T. Egan, D. Khabashesku, A. Schultz, H. Peng, V. Khabashesku, H. Furutani, K. Prather, H. Wang, S. Jackson, A. Woods, *Anal. Chem.* 76, 6734(2004).
[20] S. Ren, Y. Guo, *Rapid Commun. Mass Spectrom.* 19, 255(2005).
[21] S. Ren, Y. Guo, *J Am Soc Mass Spectrom.* 17,1023(2006).
[22] C. Wang, J. Li, S. Yao, Y. Guo, X. Xia, *Analytica Chimica Acta* 604,158(2007).
[23] Y. Gholipour, H. Nonami, R. Erra-Balsells, *J. Am Soc. Mass Spectrom.* 19, 1841 (2008).
[24] Z. Szabo, R. Vallant, A. Takatsy, R. Bakry, M. Najam-ul-Haq, M. Rainer, C. Huck, G. Bonn, *J. Mass Spectrom.* 45, 545(2010).
[25] S. Chen, H. Zheng, J. Wang, J. Hou, Q. He. H. Liu, C. Xiong, X. Kong, Z. Nie, *Anal. Chem.* 85, 6646 (2013).
[26] X. L. Kong, P. Cheng, *Materials* 3, 1845 (2010).
[27] Q. Liu, J. Shi, G. Jiang, *Trends Anal. Chem.* 37, 1 (2012).
[28] K. Tanaka, H. Waki, Y.Ido, S. Akita, Y. Yoshida, *Rapid Commun. Mass Spectrom.* 2, 151 (1988).
[29] M. Karas, F. Hillenkamp, *Anal. Chem.* 60, 2299 (1988).
[30] S. Chen, H. Zheng, J. Wang, J. Hou, Q. He, H. Liu, C. Xiong, X. Kong, Z. Nie, *Anal. Chem.*85, 6642 (2013).
[31] N. Tang, P. Tornatore, S. Weinberger, *Mass Spectrom. Rev.* 23, 34 (2004).
[32] L. A. L. Tang, J. Wang, K. P. Loh, *J. Am. Chem. Soc.* 132, 10976 (2010).
[33] X. Dong, J. Cheng, J. Li, Y. Wang, *Anal. Chem.* 82, 6208 (2010).
[34] X. Zhou, Y. Wei, Q. He, F. Boey, Q, Zhang, H. Zhang, *Chem.Commun.* 46, 6974 (2010).
[35] M. Lu, Y. Lai, G. Chen, Z. Cai, *Anal. Chem.* 83, 3161 (2011).
[36] M. Lu, Y. Lai, G. Chen, Z. Cai, *Chem.Commun.* 47, 12807 (2011).
[37] Y. Liu, J. Liu, P. Yin, M. Gao, C. Deng, X. Zhang, *J. Mass Spectrom.* 46, 804 (2011).
[38] Y. Liu, J. Liu, C. Deng, X. Zhang, *Rapid Commun. Mass Spectrom.* 25, 3223 (2011).
[39] J. Liu, Y. Liu, M. Gao, X. Zhang, *J. Am. Soc. Mass Spectrom.* 23, 1424 (2012).

[40] J. Zhang, X. Dong, J. Cheng, J. Li, Y. Wang, *J. Am. Soc. Mass Spectrom.* 22, 1294 (2011).
[41] S. Zhang, L. Ding, S. Li, X. Kong, Y. Huang, *Rapid Commun. Mass Spectrom.* 27, 1278 (2013).
[42] Q. Liu, M. Cheng, G. Jiang, *Chem. Eur. J.* 19, 5561 (2013).
[43] C.-W. Liu, M.-W. Chien, C.-Y. Su, H.-Y. Chen, .-J. Li, C.-C. lai, *Analyst* 137, 5809 (2012).
[44] Y. Liu, Y. Li, J. Liu, C. Deng, X. Zhang, *J. Am. Soc. Mass Spectrom.* 22, 2188 (2011).
[45] G. Cheng, Z.-G. Wang, Y.-L. Liu, J.-L. Zhang, D.-H, Sun, J.-Z. Ni, *Chem.Commun.* 48, 10240 (2012).
[46] D. Wan, M. Gao, Y. Wang, P. Zhang, X. Zhang, *J. Sep. Sci.* 36, 629 (2013).
[47] H. N. Abdelhamid, H.-F. Wu, *Analytica Chimica Acta* 751, 94 (2012).
[48] J. Lee, Y.-K. Kim, D.-H. Min, *J. Am. Chem. Soc.* 132, 14714 (2010).
[49] Y.-K. Kim, H.-K. Na, S.-J. Kwack, S.-R. Ryoo, Y. Lee, S. Hong, S. Hong, Y. Jeong, D.-H. Min, *ACS NANO* 5, 4550 (2011).
[50] Y.-K. Kim, D.-H. Min, *ACS Appl. Mater. Interfaces* 4, 2088 (2012).
[51] Y.-K. Kim, D.-H. Min, *Langmuir* 28, 4453 (2012).
[52] H. Kawasaki, K. Nakai, R. Arakawa, E. K. Athanassiou, R. N. Grass, W. J. Stark, Wang, *Anal. Chem.* 84, 9286 (2012).
[53] C. Shi, J. Meng, C. Deng, *Chem.Commun.* 48, 2418 (2012).
[54] B. Gulbakan, E. Yasun, M. I. Shukoor, Z. Zhu, M. You, X. Tan, H. Sanchez, D. H. Powell, H. Dai, W. Tan, *J. Am. Chem. Soc.* 132, 17408 (2010).
[55] K. Qian, L. Zhou, J. Liu, J. Yang, H. Xu, M. Yu, A. Nouwens, J. Zou, M. J. Monteiro, C. Yu, *Scientific Reports* 3, 1415 (2013).
[56] Y. –K. Kim, D.-H. Min, *Langmuir.* 25, 11302 (2009).
[57] H. Jang, Y. –K. Kim, H. –M. Kwon, W. –S. Yeo, D.-E. Kim, D.-H. Min, *Angew. Chem. Int. Ed. Engl.*49, 5703(2010).
[58] Q. Liu, J. Shi, J. Sun, T. Wang, L. Zeng, G. Jiang, *Angew. Chem. Int. Ed. Engl.* 50, 5913(2011).
[59] H. Zhang, H. K. Lee, *J. Chromatogr. A.* 1218, 4509 (2011).
[60] S. Zhang, Z. Du, G. Li, *Anal. Chem.* 83, 7531 (2011).
[61] Y.-B. Luo, J.-S. Cheng, Q. Ma, Y.-Q. Feng, J.-H. Li, *Anal. Methods* 3, 92 (2011).
[62] S. Zhang, Z. Du, G. Li, *J. Chromatogr. A.* 1260, 1 (2012).
[63] X. Wu, H. Hong, X. Liu, W. Guan, L. Meng, Y. Ye, Y. Ma, *Science of the Total Environment* 444, 224 (2013).

[64] Z. Wang, Q. Han, J. Xia, L. Xia, M. Ding, J. Tang, *J. Sep. Sci.* 36, 1834 (2013).
[65] Q. Liu, J. Shi, J. Sun, T. Wang, L. Zeng, N. Zhu, G. Jiang, *Analytica Chimica Acta* 708, 61 (2011).
[66] Q. Liu, J. Shi, T. Wang, F. Guo, L. Liu, G. Jiang, *J. Chromatogr. A.* 1257, 1 (2012).
[67] S. Tong, Q. Liu, Y. Li, W. Zhou, Q. Jia, T. Duan, *J. Chromatogr. A.* 1253, 22 (2012).
[68] J. Wu, L. Chen, P. Mao, Y. Lu, H. Wang, *J. Sep. Sci.* 35, 3586 (2012).
[69] S. Tong, X. Zhou, C. Zhou, Y. Li, W. Li, W. Zhou, Q. Jia, *Analyst* 138, 1549 (2013).
[70] J. Chen, J. Zou, J. Zeng, X. Song, J. Ji, Y. Wang, J. Ha, X. Chen, *Analytica Chimica Acta* 678, 44 (2010).
[71] Q. Liu, J. Shi, L, Zeng, T. Wang, Y. cai, G. Jiang, *J. Chromatogr. A.* 1218, 197 (2011).
[72] K.-J. Huang, Q.-S. Jing, C.-Y. Wei, Y.-Y. wu, *Spectrochimica Acta Part A* 79, 1860 (2011).
[73] K.-J. Huang, S. Yu, J. Li, Z.-W. Wu, C.-Y. Wei, *Microchim Acta* 176, 327 (2012).
[74] Y. Wang, S. Gao, X. Zang, J. Li, J. Ma, *Analytica Chimica Acta* 716, 112 (2012).
[75] H. Bao, Q. Chen, L. Zhang, G. Chen, *Analyst* 136, 5190 (2011).
[76] G. Xu, X. Chen, J. Hu, P. Yang, D. Yang, L. Wei, *Analyst* 137, 2757 (2012).
[77] J. Jiao, A. Miao, X. Zhang, Y. Cai, Y. Lu, Y. Zhang, H. Lu, *Analyst* 138, 1645 (2013).
[78] H. W. Kroto, J. R. Heath, S. C. O'Brien, R. F. Curl, R. E. Smalley, *Nature* 318, 162(1985).
[79] E. E. B. Campbell, G. Ulmer, B. Hasselberger, H. G. Busmann, I. V. Hertel, *J. Chem. Phys.* 93, 6900 (1990).
[80] P. F. Greenwood, M. G. Strachen, H. J. EI-Nakat, G. D. Willett, M. A. Wilson, M. I. Attalla, *Fuel.* 69, 257 (1990).
[81] O. Sedo, M. Alberti, J. Janca, J. Havel, *Carbon* 44, 840 (2006).
[82] X. Kong, S. Li, S. Zhang, Y. Huang, Y. Chen, *J. Am. Soc. Mass Spectrom.* 22, 2033 (2011).
[83] X. Kong, Y. Huang, Y. Chen, *J. Mass Spectrom.* 47, 523 (2012).
[84] A. Chuvilin, U. Kaiser, E. Bichoutskaia, N. A. Besley, A. N. Khlobystov, *Nature Chem.* 2, 450 (2010).
[85] S. Yang, Z. Gao, Q. Zhu, F. Kong, *Carbon* 35, 767 (1997).

[86] C. R. Wang, R. B. Huang, Z. Y. Liu, L. S. Zheng, *Chem. Phys. Lett.* 227, 109 (1994).
[87] J. M. Hunter, J. F. Fye, E. J. Roskamp, M. F. Jarrold, *J. Phys. Chem.*98, 1810(1992).
[88] S. Irle, G. Zheng, Z, Wang, K. Morokuma, *J. Phys. Chem. B* 110, 14351(2006).
[89] C. Berger, Z. Song, T. Li, X. Li, A. Y. Ogbazghi, R. Feng, Z. Dai, A. N. Marchenkov, E. H. Conard, P. N. First, W. A. de Heer, *J. Phys. Chem. B* 108, 19912(2004).
[90] S. Stankovich, D. A. Dikin, R. D. Piner, K. A. Kohlhaas, A. Kleinhammes, Y. Jia, Y. Wu, S. T. Nguyen, R. S. Ruoff, *Carbon* 45, 1558(2007).
[91] X. S. Li, W. W. Cai, J. An, S. Kim, J. Nah, D. X. Yang, R. Piner, A. Velamakanni, I. Jung, E. Tutuc, S. K. Banerjee, L. Colombo, R. S. Ruoff, *Science* 324,1312 (2009).
[92] S. Park, S. R. Rodney, *Nat. Nanotechnol.* 4, 217(2009).
[93] Y. Zhu, S. Murali, W. Cai, X. Li, J. W. Suk, J. R. Potts, R. S. Ruoff, *Adv. Mater.* 22, 3906 (2010).
[94] Z. Sun, Z. Yan, J. Yao, E. Beitler, Y. Zhu, J. M. Tour, *Nature* 468, 549(2010).
[95] L. Zhang, X. Li, Y. Huang, Y. Ma, X. Wan, Y. Chen, *Carbon* 48, 2367(2010).
[96] 96.Y. Huang, W. Yan, Y. Xu, L. Huang, Y. Chen, *Macromol. Chem. Phys.* 213, 1101(2012).

In: Graphene
Editor: Bruce T. Edwards

ISBN: 978-1-62948-795-3

Chapter 5

INCORPORATION OF GRAPHENE INTO DIRECT-PATTERNABLE TRANSPARENT CONDUCTING OXIDE THIN FILMS

Hyung-Ho Park, Hyuncheol Kim and Kyung-Mun Kang
Department of Materials Science and Engineering, Yonsei University, Seoul, South Korea

ABSTRACT

Transparent metal oxide thin films, such as zinc oxide (ZnO) and tin dioxide (SnO_2) thin films, have been increasingly studied due to their high optical transparency with electrical conducting property, high chemical durability and low cost. An incorporation of graphene into direct-patternable transparent metal oxide thin films was carried out to enhance the electrical conducting property. Metal precursors, solvent, and photosensitive additive were used for the synthesis of stock solutions. After addition of graphene, the solution was spin-coated and the coated film was exposed to ultraviolet (UV) light of 365 nm. Then, the UV exposed film was washed by solvent and the direct-patterned film was annealed to form a desired pure phase system. From the results of UV-vis spectra, X-ray diffractometer, Hall effect measurement and photoelectron spectroscopy of the graphene incorporated metal oxide thin films, we can deduce the effect of graphene in transparent metal oxide thin films.

1. INTRODUCTION

Recently, there has been rapidly changing demands for highly efficient and various displays, such as liquid crystal display, plasma display panel, organic light-emitting displays (LEDs), LED television (TV) and 3-dimensional TV. In addition to displays, transparent electronic materials have also been essential in solar cells [1,2]. Materials used as a transparent electrode should have optical transparency greater than 80% in the visible light region (380-780 nm) and high electrical conductivity (>10^3 $\Omega^{-1}cm^{-1}$). Indium tin oxide (ITO) has good electrical conductivity and productivity, therefore, it has been widely used in transparent electrodes [3,4]. However, there are many disadvantages of ITO, including the high cost due to the scarcity of indium, instability in the hydrogen plasma, poor mechanical properties and the use of a toxic etching solution [5,6]. Zinc oxide (ZnO) and tin dioxide (SnO_2) thin films have been increasingly studied in recent years as transparent electrode for display devices and gas sensors due to their high optical transparency, high chemical durability, and low cost as a starting material [7,8].

1.1. Transparent Conducting Oxide

The simultaneous occurrence of high optical transparency (more than 80%) in the visible region and high electrical conductivity (> 10^3 $\Omega^{-1}cm^{-1}$) is not possible in an intrinsic stoichiometric material. Partial transparency and fairly good conductivity may be obtained in thin films of a variety of metals. The only way to obtain good transparent conductors is to create electron degeneracy in a wide band gap (> 3 eV) oxide by controllably introducing non-stoichiometry and/or appropriate dopants. Transparent conducting oxide (TCO) is achieved by selecting a wide-bandgap oxide that is rendered degenerate through the introduction of native or substitutional dopants. Numerous techniques for depositing several very useful transparent conducting oxides, notably antimony-doped tin oxide or fluorine-doped tin oxide and ITO have been developed, some at a large-scale production level. These thin film devices include the following: resistors; transparent heating elements for aircraft and automobile windows; antistatic coatings for instrument windows; heat-reflecting mirrors for glass windows and incandescent bulbs; antireflection coatings; selective absorber components in solar heat collectors; gas sensors; electrodes for liquid crystal, electrochromic and ferroelectric photoconductor storage and display devices; amorphous hydrogenated silicon (a-Si:H) solar cells; protective and wear-resistant coatings for glass containers.

1.1.1. Zinc Oxide

Zinc oxide is a technologically important material. The electrical and optical properties of pure and also doped bulk ZnO have been studied [9,10]. Thin films of ZnO have been prepared by various techniques [11,12]. ZnO films retain the bulk wurtzite structure as shown in Figure 1 and are composed of columnar crystallites with grain sizes [13] in the range 50-300 Å. Sputtered ZnO films have a strong c axis orientation, perpendicular or parallel to the substrate depending mainly on the substrate material, while the degree of preferred orientation depends on the deposition parameters [14,15]. ZnO shows a relatively large direct band gap of ~ 3.3 eV. Undoped ZnO films deposited using r.f. magnetron sputtering in an Ar-H_2 mixture at $T \approx 75°C$ had a typical as-deposited resistivity of about 2 x 10^{-3} Ω·cm ($N \approx 10^{20}$ cm^{-3}; $\mu \approx 8$ $cm^2/V \cdot s$) and $T \approx 90\%$ [16]. ZnO shows mainly *n*-type character. Nonstoichiometry is an origin of *n*-type character, but the subject remains controversial [17]. Controllable n-type doping is easily achieved by substituting Zn with group-III elements and by substituting oxygen with group-VII elements [18].

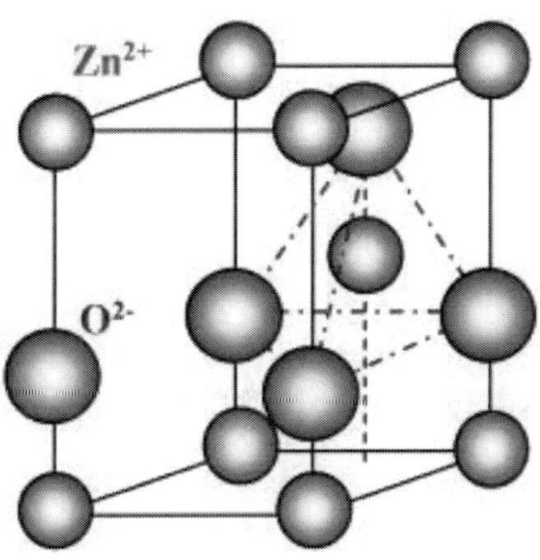

Figure 1. Wurtzite structure of ZnO.

1.1.2. Tin Oxide

Under optimum conditions of deposition undoped SnO_2 films are generally polycrystalline (grain size, about 200 ~ 300 Å [19-21]) and retain the tetragonal rutile structure of bulk SnO_2 as shown in Figure 2 [22]. The existence of a preferred orientation of the crystallites and its dependence on substrate temperature [23] and on oxygen partial pressure in the case of sputtering [24] is well established, although reports on the exact nature of the orientation lack consistency. As already mentioned, the presence of the tin oxide (SnO) phase and the stoichiometry [25] in reactively sputtered films depend on several deposition parameters. For films prepared by chemical vapor deposition or by spraying, the reports on the presence of the SnO phase are quite conflicting [26,27].

The n-type conductivity of undoped SnO_2 is primarily due to its nonstoichiometry, but in films deposited from chlorides the incorporation of chlorine ions into the lattice [28] also contributes to the conductivity. SnO_2 films are degenerate semiconductors, typically with a carrier concentration $N \approx 10^{19}$-10^{20} cm^{-3}, a mobility $\mu \approx$ 5-30 cm^2/V·s and a resistivity $\rho \approx 10^{-3}$-10^{-2} Ω·cm (corresponding to a sheet resistance of about 100-500 Ω/sq). Annealing of SnO_2 films at 400°C in oxidizing or reducing ambients causes a large change in μ while N remains weakly affected. Pure SnO_2 films exhibit a direct optical band gap of 3.87-4.3 eV as reported by many researchers [29-31]. The visible and near-IR transmittance is high (about 80%). The refractive index is fairly constant in the visible region and ranges typically between 1.8 and 2.0 [32,33]. The hardness of SnO_2 films on Moh's scale is 7-8 (as for topaz) [34]. The films are stable under typical environmental conditions and resistant to chemical etching. Their adhesion to glass or porcelain is very high (about 200 kgf cm^{-2}) [35]; hence the film is possibly attached to the substrate by chemical bonds.

1.2. Direct-Patterning of Functional Films

For the fabrication of electronic circuits, micro-patterning is a commonly used process. However, the conventional dry etching process is accompanied by the generation of physical defects, degradation of properties, and pollution from hazardous materials. To address these problems, photochemical solution deposition by using photosensitive precursor or additive can be used for direct-patterning of films [36-38]. If direct-patterning process is used, photoresist and dry etching are not necessary for microscale patterning, as the coated films behave like the negative photoresist. Thus, damage and problems related to dry etching can be avoided. Process steps of conventional etching and direct-patterning process were shown in Figure 3. Left is a general process step of conventional etching process and right is direct-pattering process. Direct-patterning has short steps when comparing with conventional etching. Ultraviolet (UV) exposed photosensitive solution changes to pattern and unexposed area is washed away by solvent. Finally, direct-patterned functional film can be obtained.

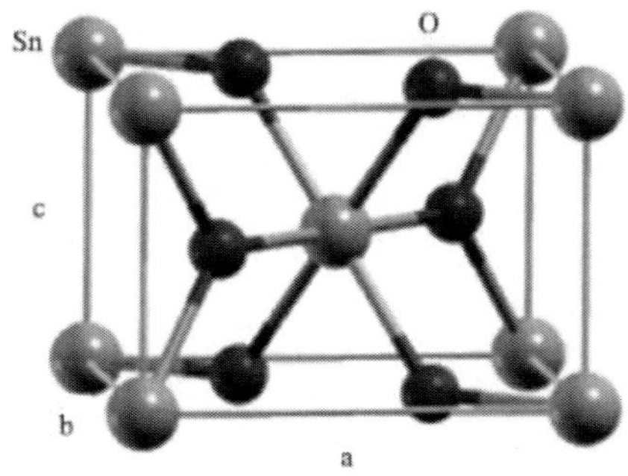

Figure 2. Rutile tetragonal structure of SnO_2.

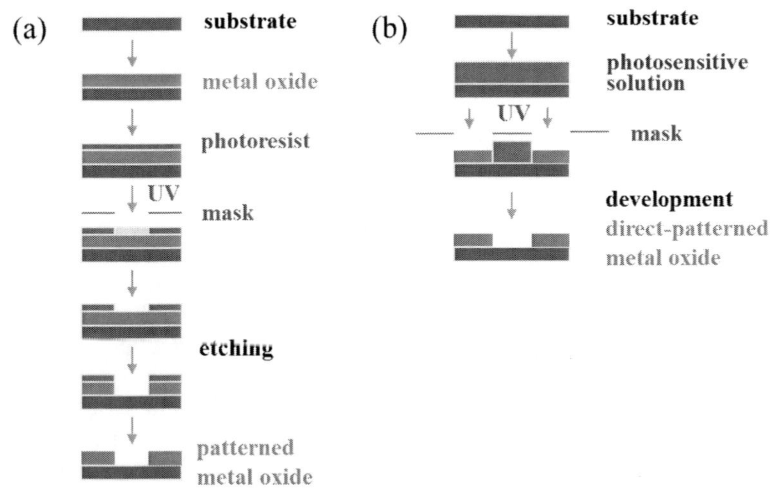

Figure 3. Comparison between (a) conventional etching process and (b) direct-patterning process steps.

A micro-patterning process for the fabrication of electronic circuits has been developed. However, the conventional dry etching process is accompanied by the generation of physical defects, degradation of the properties, and pollution from hazardous materials. To address these issues, photochemical solution deposition has been used for direct-patterning of thin films, which eliminates the use of a photoresist and dry etching for micron-scale patterning as the coated films behave like a negative photoresist [36,39]. As a result, damage and problems caused by dry etching can be avoided.

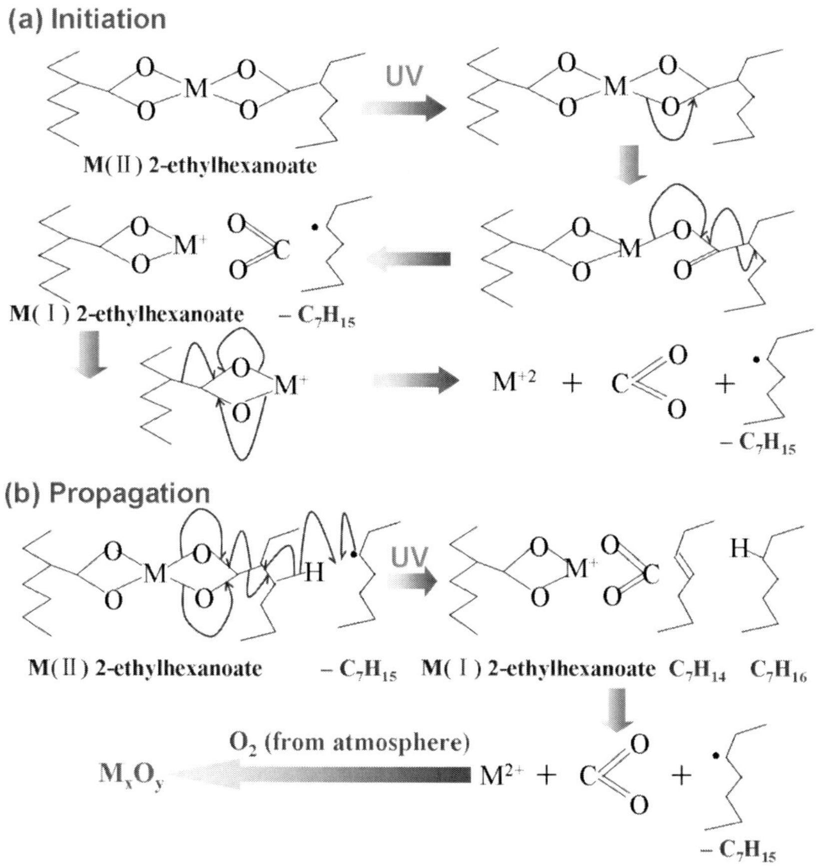

Figure 4. (a) Initiation and (b) propagation of M(II) 2-ethylhexanoate during UV exposure.

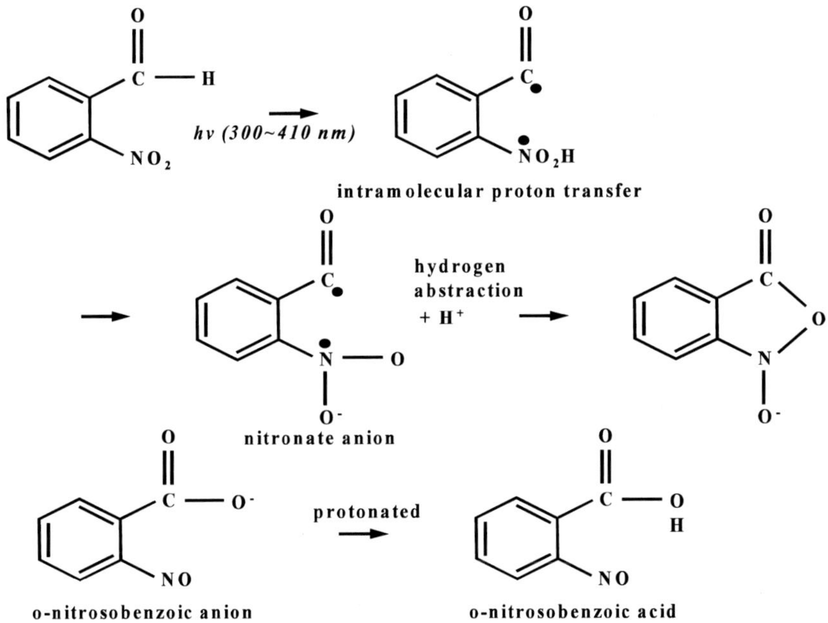

Figure 5. A photodissociation reaction of o-nitrobenzaldehyde (NBAL).

2. Formation Mechanism of Direct-Patterning of TCO Films

2.1. Direct-Patterning by Photochemical Metal-Organic Deposition

A photochemical reaction is investigated using the precursors with 2-ethylhexanoate ligand. As shown in Figure 4(a), an initiation step, during UV-irradiation fragmentations of metal 2-ethylhexanoate are happened by ligand-to-metal charge transfer [40-42]. As a result of charge transfer, C_7H_{15} radical and carbon dioxide (CO_2) are generated with positive ionic state of metal. Furthermore, charge transfer induces the generation of second radical, CO_2, and metal ion [40]. If this metal ion reacts with oxygen in the atmosphere, metal oxide can be prepared. As shown in Figure 4(b), a propagation step, the other type of a photochemical reaction, hydrogen-abstraction of metal 2-ethylhexanoate is restarted with hydrogen-abstraction from the coordinated 2-ethylhexanoate ligand and resulted in a radical chain process [40,41].

Furthermore, charge transfer induces the generation of second radical, CO_2 and metal ion. If we control the atmosphere of a photochemical reaction, we can obtain the metal as metallic form or oxide form.

2.2. Direct-Patterning by Photochemical Solution Deposition

The photolysis reaction stage of the photosensitive additive is explained as follows. The metal ion in metal-alkoxide is an acid and aldehyde is a base, resulting in a strong attraction that limits hydrolysis and condensation reactions. After UV exposure, intermolecular proton transfer and hydrogen abstraction occur, and a 2-nitrosobenzoic anion is produced (Figure 5 [43]). The metal ion in the metal-alkoxide formed by the condensation reaction attracts the nitryl ligand and a cross-linked structure is formed (Figure 6 [43]). This cross-linked structure is no more fluidic state and can't be removed by dissolution in the following solvent-rinsing step [44].

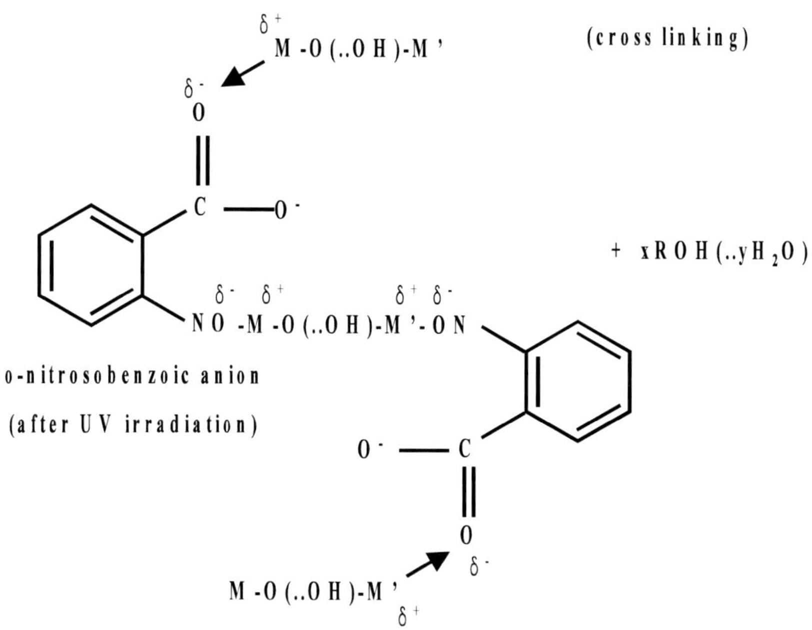

Figure 6. Cross-linking mechanism of photoreaction with NBAL and metal alkoxide.

3. Fabrication of Graphene-Incorporated Direct-Patternable Metal Oxide Thin Films

3.1. Fabrication of Graphene-Incorporated Direct-Patternable ZnO Thin Film

3.1.1. Experimental

ZnO thin films were prepared through a sol–gel process. Zinc acetate dihydrate ($Zn(CH_3COO)_2 \cdot 2H_2O$), 2-methoxyethanol, and monoethanolamine (MEA) were used as the starting Zn precursor, solvent, and sol stabilizer, respectively. Zinc acetate dihydrate was dissolved in 2-methoxyethanol, MEA was added for stabilization at room temperature, and NBAL was introduced as a photosensitive additive. Graphene nanopowder (AO-2, Graphene supermarket) was used as a graphene source. The thickness and lateral size of graphene were approximately 8 nm and 200 nm, respectively. The concentration of zinc acetate dihydrate was 0.5 M, and 0.012, 0.025, 0.05 and 0.1 wt.% graphene were incorporated into the ZnO photosensitive solution. The mixed solution was stirred at room temperature for 3 h. The graphene in mixed solution was dispersed by a sonicator and then spin-coated onto Si and glass (Corning 1737) at 2000 rpm for 20 s. For direct-patterning of the film, the spin-coated film was exposed to 365 nm UV light for 10 min. The UV-exposed film was washed with 2-methoxyethanol to remove the unexposed area. After washing, the films were dried at 250°C for 5 min on a hot plate to remove the solvent and organic residuals. This spin-coating and preheating procedure was repeated two times to obtain the desired thickness. After the final layer deposition, the film was annealed in a tube furnace under a N_2 atmosphere at 500°C for 1 h. The crystallinity was analyzed using an X-ray diffractometer (XRD, D/MAX-2000, Rigaku) with Cu Ka radiation. Scan was performed using theta/2theta ($\theta/2\theta$) method in the procedure of 0.05° step with a 4°/min scan speed. The final thickness of the film was measured using ellipsometry (L117, Gaertner) with a He-Ne laser source (λ = 632.8 nm). The electrical properties of the film were measured using Hall effect measurements (HMS-3000, Ecopia). Optical transmittance measurements were performed using a UV-vis-NIR spectrophotometer (V-570, Jasco). Scanning electron microscopy (SEM, JSM-6390, JEOL) was used to investigate the film microstructure with an operating voltage of 15 kV. X-ray photoelectron spectroscopy (XPS, ESCALAB 220i-XL, VG Scientific) with an Al Kα monochromatic source was used to determine the composition and surface chemical bonding state of the film. The accelerating voltage and emission current of X-ray source were 15 kV and 20 mA, respectively. The atomic percentage was calculated from core level XPS spectra using atomic sensitivity factors [45]. The binding energies were corrected using the reference C 1s peak at 284.5 eV [45]. A Shirley subtraction and Gaussian shape were used for background subtraction and peak fit analysis, respectively. Avantage 3.25 was used for peak fitting software supplied by XPS manufacturer, VG Scientific. During peak fit analysis of measured spectra, the parameters such as binding energy and full width at half maximum values were constrained refer to the published references [46-52].

3.1.2. Results and Discussion

The direct-patterning of graphene-incorporated ZnO thin film was examined by removing the area not exposed to UV using 2-methoxyethanol. The SEM image in Figure 7 indicates a relatively bright area corresponding to the graphene-incorporated ZnO thin film and a dark area corresponding to the Si substrate [53]. As shown at the pattern edge, patterning with a scale of several tens of microns was successfully obtained by lithography using photochemical solution deposition.

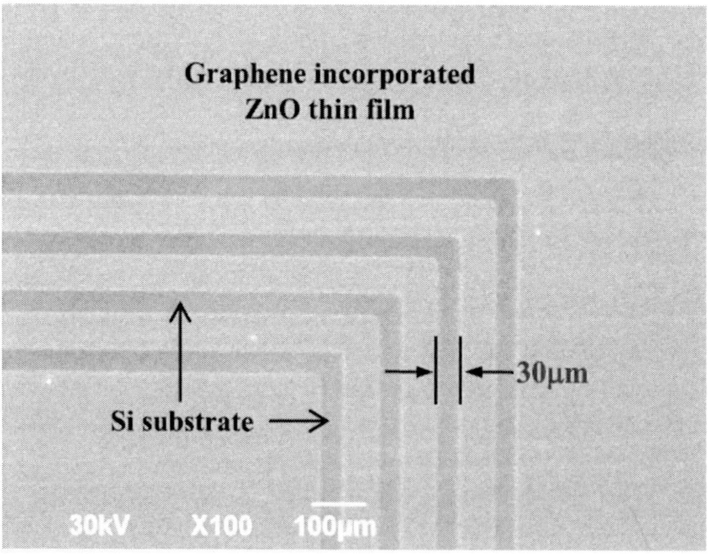

Figure 7. Optical image of direct-patterned, graphene-incorporated ZnO thin film dried at 250 °C for 5 min.

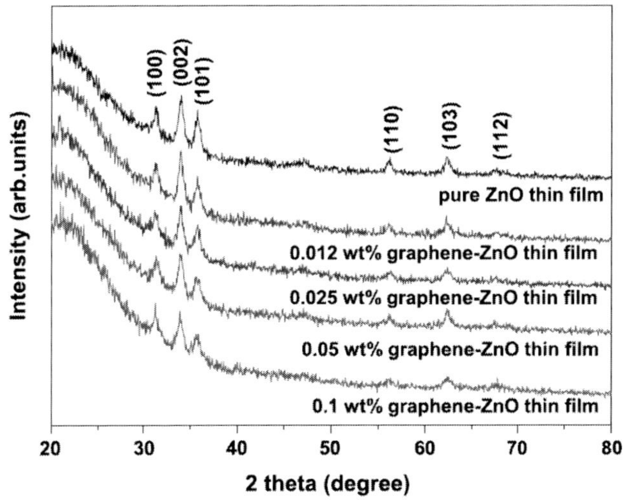

Figure 8. XRD patterns of ZnO thin films containing various amounts of graphene after annealing at 500°C under a N_2 atmosphere.

Figure 8 presents XRD patterns of ZnO thin films containing various amounts of graphene annealed at 500°C under a N_2 atmosphere. Based on the literature and previous experiments, 500°C was a sufficient annealing temperature for ZnO phase formation [53-56]. All the films were indexed with (100), (002), (101), (110), (103) and (112) diffraction peaks of the ZnO hexagonal wurtzite crystalline phase [57]. The intensity of the diffraction pattern was affected by the film thickness as measured by ellipsometry, which was set at approximately 70 nm. There was no significant difference between the thickness of ZnO and those of varying contents of graphene-incorporated ZnO thin films. When graphene was incorporated, similar diffraction patterns were observed as that of pristine ZnO thin film, as shown in Figure 9. However, the presence of graphene did not significantly affect the crystallinity of the films, and the diffraction pattern intensity decreased only slightly with an increase in graphene content. Under these experimental conditions, the crystallinity and growth orientation of ZnO thin films containing graphene did not change, indicating that graphene does not affect the crystallization of ZnO thin films.

The electrical properties of pure ZnO and graphene-incorporated ZnO thin films annealed at 500°C under a N_2 atmosphere were measured using Hall effect measurements, and the results are shown in Figure 9 [53]. The values confirm that the carrier concentration, mobility, and resistivity were affected by the content of the incorporated graphene. The carrier concentration and mobility of pure ZnO thin film were 9.74 x 10^{18} cm^{-3} and 0.26 $cm^2/V{\cdot}s$, respectively. The carrier concentration slightly decreased with increasing graphene incorporation. This carrier concentration behavior on ZnO thin films was caused by the carrier concentration of graphene. The theoretical electron carrier concentration of graphene was nearly 10^{12} e/cm^2 [58]. Therefore, the carrier concentration gradually decreased with increasing graphene incorporation. The mobility increased with an increase in graphene content. The improved mobility was related to the π-bonds on the surface and inner-shell of the graphene flakes in the ZnO matrix. The π-bonds closest to the Fermi level are weak, allowing easy movement of the charge carrier [59,60]. The variation in resistivity is proportional to the carrier concentration and mobility, through which the resistivity of graphene-incorporated ZnO thin films decreased with an increase in the content of incorporated graphene [61]. The carrier concentration and mobility of 0.1 wt.% graphene-incorporated ZnO thin film were 2.41 x 10^{18} cm^{-3} and 1.76 $cm^2/V{\cdot}s$, respectively.

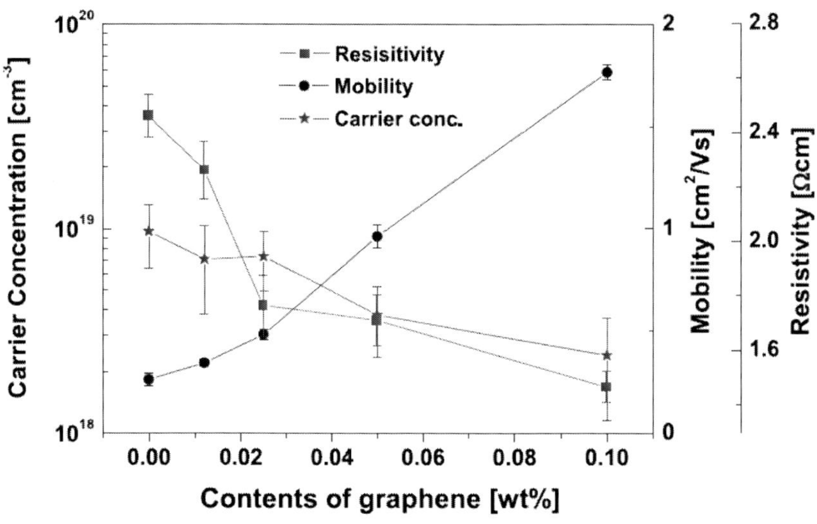

Figure 9. Carrier concentration (★), mobility (●), and resistivity (■) of graphene-incorporated ZnO thin films after annealing at 500 °C under a N_2 atmosphere.

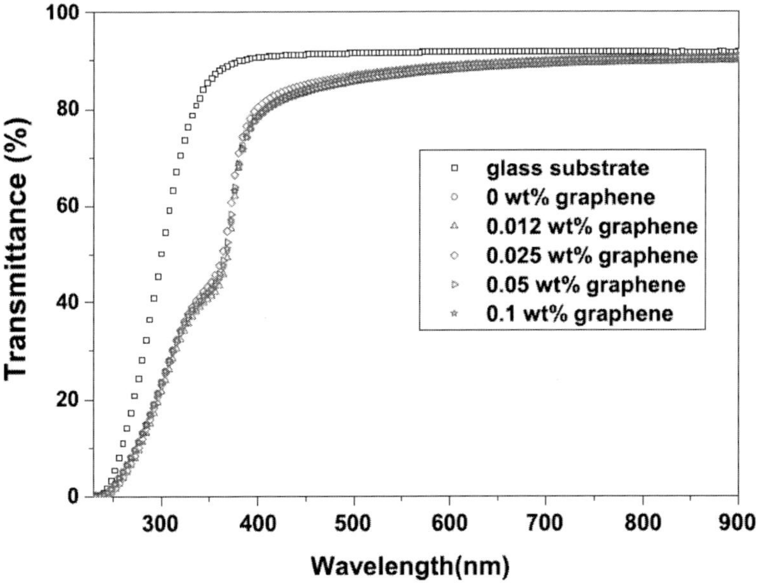

Figure 10. Optical transmittance of graphene-incorporated ZnO thin films after annealing at 500 °C under a N_2 atmosphere.

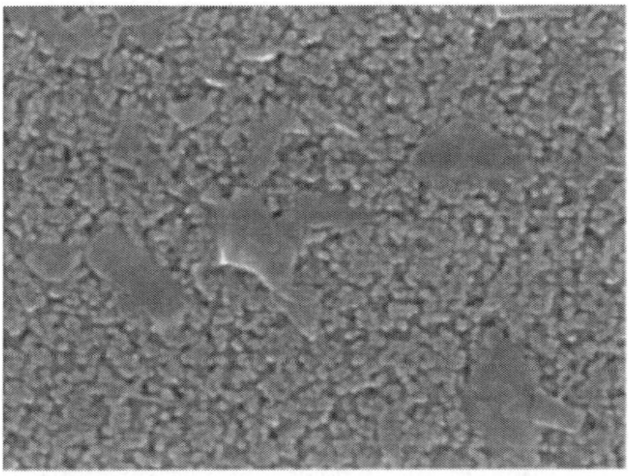

Figure 11. Planar SEM image of 0.1 wt.% graphene-incorporated ZnO thin film.

The UV-vis absorption spectra of pure ZnO and graphene-incorporated ZnO thin films are shown in Figure 10 [54]. The transmittance of the ZnO thin films was similar to those containing graphene; in the visible light region, the transmittance values of the films containing 0, 0.012, 0.025, 0.05 and 0.1 wt.% graphene were 86.9, 86.1, 87.3, 86.3 and 87.0%, respectively. Graphene did not decompose in the ZnO stock solution and therefore was present in the ZnO thin films. However, graphene did not affect the transmittance of the thin film because of its very small size, and the optical absorption values of graphene were not changed by varying wavelengths [62-64]. The planar SEM image of graphene-incorporated ZnO thin film was shown in Figure 11 [54]. From the SEM image, the major axis length of graphene was found from several tens nm to approximately 200 nm. Specifically, there was no significant difference in transmittance between the ZnO films.

The pristine graphene, ZnO, and 0.1 wt.% graphene-incorporated ZnO thin films were analyzed by XPS to determine the electrical bonding state of graphene within the ZnO matrix, and the results are given in Figure 13. The ZnO thin films were annealed at 500°C for 1 hr. under a N_2 atmosphere, and the C 1s, O 1s, and Zn 2p core levels were obtained after removing the surface contaminations of the film by Ar ion-etching in a ultra-high vacuum chamber. As shown in Figure 12(a), pristine graphene has a C-C peak at 284.5 eV and a C-OH peak at 285.6 eV [65]. Based on the C 1s core level spectra of graphene, the graphene in this experimental condition was not oxidized. The atomic percentages of C, O, and Zn in 0.1 wt.% graphene-incorporated ZnO thin films were 4.9, 30.2, and 64.9 at%, respectively. The value of 4.9 at% corresponds to the amount of total carbon in the ZnO thin film. A 0.7 at% content of graphene in graphene-incorporated ZnO thin film was calculated from C-C peaks at 284.5 eV. However, because the surface contaminated carbon was already removed by ion etching, there was no C-C peak for graphene in the C 1s spectra of the ZnO thin film. The other carbon peaks were attributed to the presence of loosely bound carbons in C-OH and carboxyls bonds on the ZnO matrix [44,45]. The carboxyl bonds originated from the NBAL photosensitive additive because it is not easily decomposed at temperatures lower than 600°C [47]. In Figure 12(b), the O 1s peaks of ZnO and graphene-incorporated ZnO thin films were fitted, and the binding energy peaks were located at 530.2, 530.8 and 532.2 eV, respectively [54]. The peak binding energy peaks near 530.2 and 530.8 eV were attributed to oxygen in O-Zn bonds and oxygen-deficiencies, respectively. The other peak positioned at 532.2 eV was ascribed to hydroxyl groups, -OH, on the surface of the ZnO structure [48,49]. The deficient peak of graphene-incorporated ZnO was slightly greater than that in the ZnO thin film due to an increase in surface area due to graphene incorporation [50]. The increased surface area by graphene incorporation makes more oxygen deficient state than pristine ZnO film. As shown in Figure 12(c), the Zn 2p core level spectra of 0.1 wt.% graphene-incorporated ZnO film was identical to those of the ZnO film. Changes in the core level spectra of Zn 2p indicate a chemical bond between graphene and the ZnO matrix. Since no significant shift was observed in the peak position of the Zn 2p spectra, there was no chemical Zn-C and/or Zn-O-C bond between graphene and the ZnO matrix under our experimental conditions [50-52]. Therefore, the electron transfer of graphene in the ZnO matrix or vice-versa was maintained by the π-bonds on the graphene surface, with no chemical bonds occurring between the graphene and ZnO matrix [66]. Therefore, as shown in Figure 3, enhanced resistivity after the incorporation of graphene into the ZnO film is a result of enhanced charge mobility due to the π-bonds of graphene.

3.2. Fabrication of the Graphene Incorporated Direct-Patternable SnO_2 Thin Film

3.2.1. Experimental Details

SnO_2 thin films were prepared by a photochemical solution deposition. Precursor for the tin source and solvent were tin chloride pentahydrate ($SnCl_4{\cdot}5H_2O$) and methanol (CH_3OH), respectively. $SnCl_4{\cdot}5H_2O$ was dissolved in CH_3OH at a final concentration of 0.3 M and NBAL was introduced as a photosensitive additive. The dissolved photosensitive solution was stirred at room temperature for 3 hr. Graphene sheets were micromechanically cleaved from highly oriented pyrolytic graphite [67]. Graphene in the mixed solution were dispersed by a sonicator [68]. We could obtain a solution for graphene incorporated SnO_2 thin film with containing various weight ratios. The weight ratio was controlled from 0 to 0.1. The solution was spin-coated at 2,000 rpm for 20 sec on corning 1737 glass substrates. The spin-coated film was dried on hot plate at 60°C and the dried film was exposed to UV light of 365 nm. Then, the UV exposed film was washed by CH_3OH to remove the unexposed area. And the films were dried at 150°C for 10 min and annealed at 500°C for 1 hr in a tubular furnace under an N_2 atmosphere [69]. Thickness of the films was obtained by using ellipsometry. The crystallinity was analyzed by using X-ray diffractometer (Ultima IV, Rigaku) with Cu Kα radiation. The electrical properties of the films were measured using a Hall effect measurement (HMS-3000, Ecopia). Optical transmittance measurements were carried out using a UV-vis-NIR spectrophotometer (V-570, Jasco). To investigate the surface chemical bonding state, photoelectron spectroscopy (PES) was performed by photon energy at 630 eV and 130 eV in the 8A2 beam line of the Pohang Accelerator Laboratory.

3.2.2. Results and Discussion

Figure 13 presents XRD patterns of SnO_2 thin films containing various amounts of graphene annealed at 500°C. All the films were indexed as (110), (101), (200), and (211) diffraction peaks of the cassiterite SnO_2 crystalline phase [70,71]. When the graphene was incorporated, similar diffraction patterns were observed. The presence of graphene did not affect the crystallinity of the films significantly and the diffraction pattern intensity decreased only slightly with an increase of graphene content. With our experimental conditions, the crystallinity and growth orientation of SnO_2 thin films containing graphene were not changed. In other words, graphene do not affect the crystallization of SnO_2 thin films.

UV-vis absorption spectra of SnO_2 and graphene incorporated SnO_2 thin films were shown in Figure 14 [71]. As shown in Figure 14, the transmittance of SnO_2 thin films was very slightly decreased by incorporation of graphene. At visible light region, the transmittance values of the films with containing 0, 0.01, 0.02, 0.05 and 0.1 wt.% graphene were 90.9, 90.8, 90.8, 90.7 and 90.6%, respectively. A slight decrease in transmittance occurred with the incorporation of graphene because graphene act as a scattering center in the film. Because graphene are not decomposed in the SnO_2 stock solution, graphene exist in SnO_2 thin films [72].

The electrical properties of pure SnO_2 and graphene incorporated SnO_2 thin films were measured by using a Hall Effect measurement and the results are given in Figure 15 [71]. The values confirm that the carrier concentration, mobility, and resistivity were affected by contents of incorporated graphene. The carrier concentration and mobility of SnO_2 thin film appeared to 5.44 x 10^{18} cm^{-3} and 2.78 $cm^2/V \cdot s$, respectively. An increase in the mobility was larger than an increase in the carrier concentration with graphene incorporation, for example, more than 5 times for the mobility but less than 2 times for the carrier concentration. The improved behaviour of mobility is related with π-bond on surface of graphene in SnO_2 matrix. The π-bond is weak bond that they lie closest to the Fermi level. Due to the π-bond, charge carrier can move easily on the surface of graphene [73,74]. A relative small increase in the carrier concentration is because graphene does not act as dopants which can generate free carriers [75]. The variation of the resistivity is proportional to the carrier concentration and mobility. By the carrier concentration and mobility, the resistivity of graphene incorporated SnO_2 thin films was decreased with an increase of graphene [62]. The carrier concentration and mobility of 0.1 wt.% graphene incorporated SnO_2 thin film appeared to 4.013 x10^{18} cm^{-3} and 16.13 $cm^2/V \cdot s$, respectively.

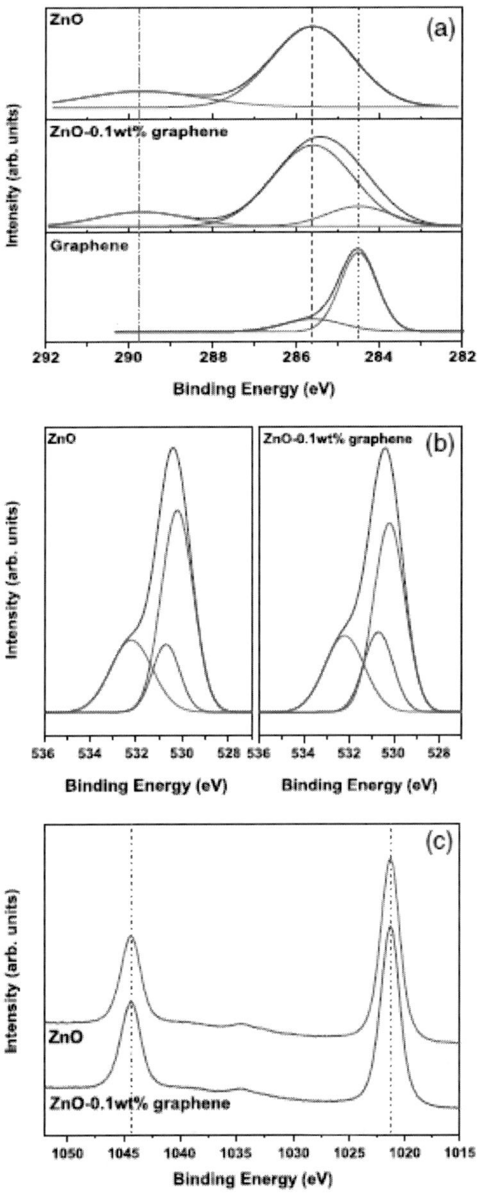

Figure 12. XPS spectra of (a) C 1s, (b) O 1s, and (c) Zn 2p core levels of ZnO and 0.1 wt.% graphene-incorporated ZnO thin films after annealing at 500°C under a N_2 atmosphere.

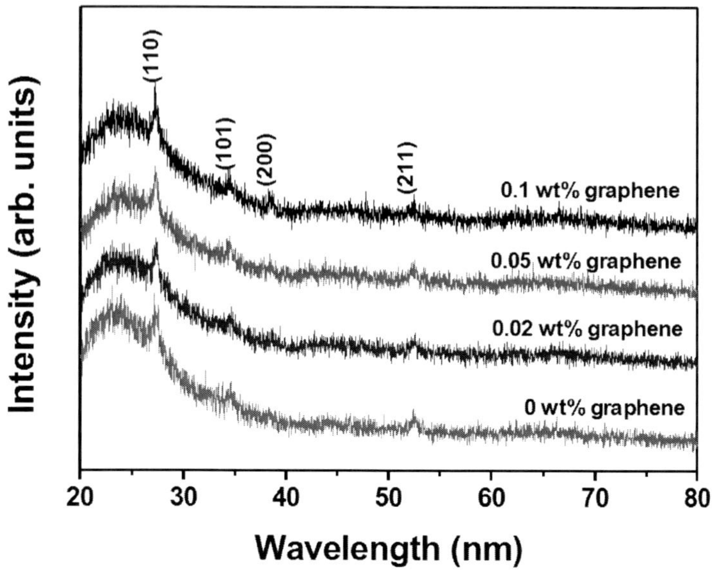

Figure 13. XRD patterns of SnO_2 and graphene incorporated SnO_2 thin films annealed at 500°C.

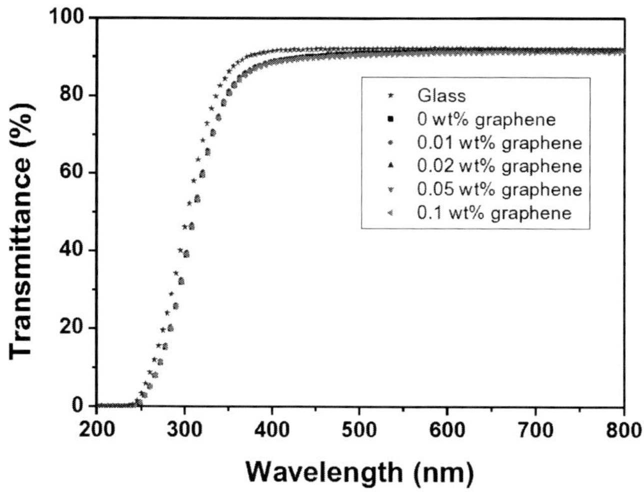

Figure 14. Optical transmittance of SnO_2 and graphene incorporated SnO_2 thin films annealed at 500°C.

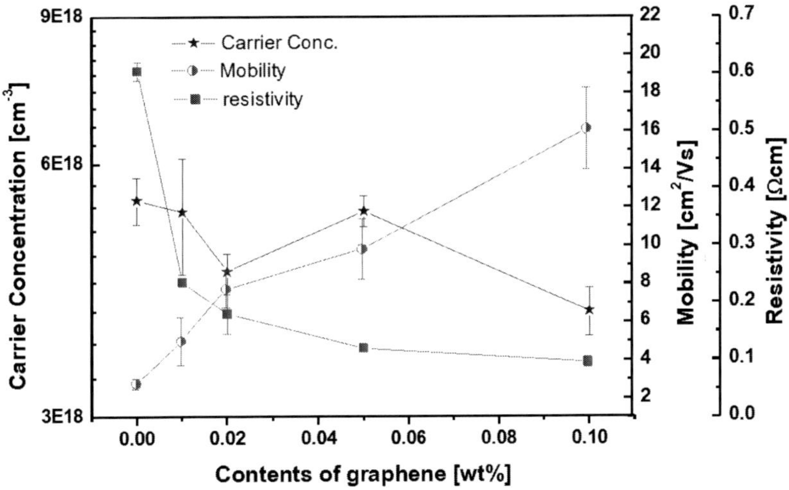

Figure 15. Carrier concentration (★), mobility (◑), resistivity (■) of SnO_2 and graphene incorporated SnO_2 thin films annealed at 500°C.

The pure SnO_2 and 0.1 wt.% graphene incorporated SnO_2 thin films were analyzed by PES to determine the electrical bonding state of the graphene within the SnO_2 matrix. Both films were annealed at 500°C for 1 hr and the C 1s, O 1s, Sn 3d core levels, and valence band spectra were measured. The binding energies were corrected using the referencing C 1s peak at 284.5 eV [45]. Figure 16 shows (a) Sn 3d core level spectra and (b) valence band spectra of pure SnO_2 and 0.1 wt.% graphene incorporated SnO_2 hybrid films [71]. Changes of core level spectra of Sn 3d would indicate a chemical bond between graphene and SnO_2 matrix. However, as shown Figure 16(a), Sn 3d core level spectra of 0.1 wt.% graphene incorporated SnO_2 hybrid film was identical to those of pure SnO_2 film. This indicates that there was no chemical bond between graphene and SnO_2 matrix under our experimental conditions [71,76]. Therefore, we can infer that electron transfer of graphene in SnO_2 matrix or vice-versa was maintained by π-bond in the surface of graphene without any chemical bond between graphene and SnO_2 matrix [66]. In Figure 16(b), valence band maximum energy (E_{VBM}) of 0.1 wt.% graphene incorporated SnO_2 film was identical to that of pure SnO_2 film [71]. The level of E_{VBM} was related with density of electrons on energy band diagram of film [77]. Therefore, as shown in Figure 16, enhanced resistivity and similar carrier concentration after incorporation of graphene into SnO_2 thin film is a result of an enhancement of charge mobility due to π-bond of graphene [71].

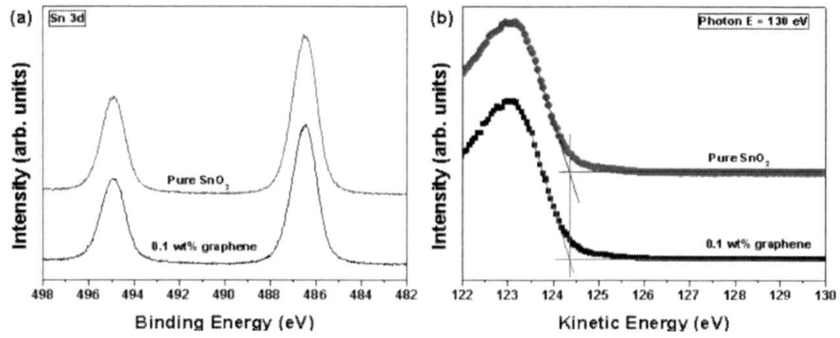

Figure 16. PES spectra of (a) Sn 3d core level and (b) valence band maximum region of SnO_2 and 0.1 wt% graphene incorporated SnO_2 thin films annealed at 500°C.

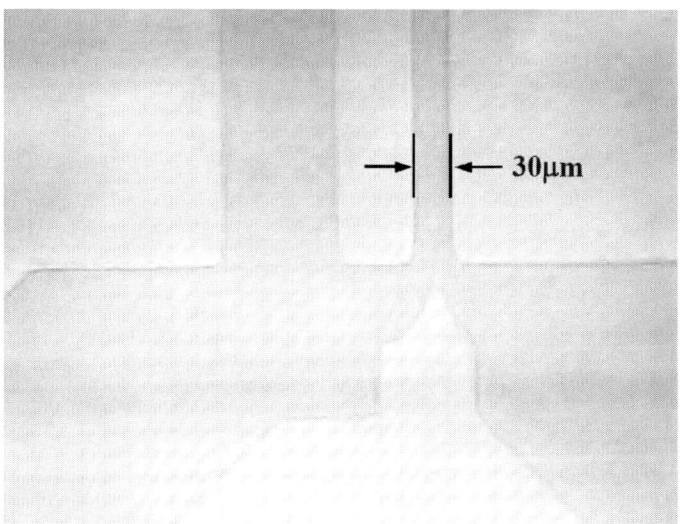

Figure 17. Optical image of direct-patterned graphene incorporated SnO_2 thin film.

The possibility of direct-patterning of graphene incorporated SnO_2 thin film was examined by removing the area unexposed to UV using a CH_3OH. The optical micrograph of Figure 17 indicates a relatively bright area corresponding to the graphene incorporated SnO_2 thin film and dark area corresponding to the glass substrate [71]. As shown at the edge of the pattern, several tens micron-scaled patterning was successfully obtained by lithographic patterning using photochemical solution deposition.

4 . Gas Sensor Application

Metal oxide semiconductor based gas sensors have attracted considerable attention owing to their low-cost, high sensitivity, and high compatibility with microelectronic processing. The gas sensing properties enhancement of TCO film by incorporation of graphene was studied for application.

A gas sensor for sensing carbon monoxide (CO) gas was fabricated using a SiO_2/Si substrate with Pt interdigitated electrodes (IDE) in which the gap between each electrode was 5 μm. The thickness of Pt was 200 nm and the IDE patterns were fabricated using photolithography and dry etching. The responses of the fabricated gas sensors to CO gas were measured at 400°C by monitoring the change in sensor resistances while changing the flow gas from dry air to test gases (100 ppm CO balanced with dry air). To eliminate interfering effects, we used a constant flow rate of 1000 sccm for dry air and the test gases. The experimental setup for gas sensing measurements is given in Figure 18 [78]. The film resistance was measured under a direct current bias voltage of 3 V using a source measurement unit (Keithley 2635a).

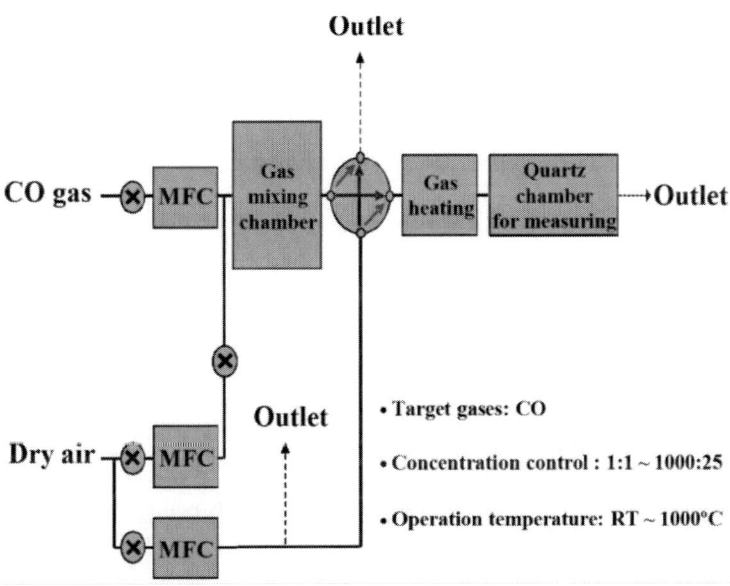

Figure 18. Schematic diagram of the system used to test sensor response. CO gas or dry air was injected into the quartz chamber through heating furnace. There, changes in sensor resistance were monitored.

4.1. CO Gas Sensing Properties of Direct-Patternable SnO_2 Films Containing Graphene

The effects by incorporation of graphene on the CO gas sensing properties of SnO_2 thin films were investigated. Through incorporation of graphene, the gas sensitivity of SnO_2 thin film gas sensors was enhanced.

The photosensitive precursor for the production of SnO_2 films was tin 2-ethylhexanoate, $Sn(O_2CCH(C_2H_5)C_4H_9)_2$. The solvent and sol stabilizer were 4-methyl-2-pentanone and monoethanolamine, respectively. Tin 2-ethylhexanoate was dissolved in 4-methyl-2-pentanone at 0.3 M and monoethanolamine was added for stabilization at a molar ratio of 1.0. The dissolved photosensitive solution was stirred at room temperature. Graphene sheets were micromechanically cleaved from highly oriented pyrolytic graphite [78]. Graphene in the mixed solution were dispersed by a sonicator [66]. To obtain SnO_2 thin films containing graphene, a 0.05 wt.% graphene was incorporated into the SnO_2 photosensitive solution. The content of graphene was selected by considering of atomic weight and percolation effect from previous results [36,78]. The solution was spin-coated at 2000 rpm for 30 s on glass and Si substrates with Pt IDE. For direct-patterning, the spin-coated film was exposed to UV light with a 365 nm wavelength. Then, the UV exposed films were washed with 4-methyl-2-pentanone to remove the unexposed area of the film. After washing, the films were aged at 50°C for 12 h in a dry oven and annealed at 500°C for 1 h in a tubular furnace under an O_2 atmosphere to remove the solvent and organic residues. Finally, the films were annealed in a vacuum furnace at 450°C for 12 h.

XRD patterns of SnO_2 films containing graphene annealed at a temperature of 500ºC are presented in Figure 19 [79]. The patterns of the films were indexed as (110), (101), (200), (211) and (220) diffraction peaks of cassiterite crystalline phase SnO_2 [70]. Similar diffraction patterns were observed by the incorporation of graphene. The presence of graphene did not affect the crystallinity of the films significantly, and the diffraction pattern intensity was decreased. The presence of graphene potentially increased the impurity content in the SnO_2 thin films and may act as a grain growth inhibitor.

The gas response characteristics of gas sensors based on SnO_2 thin films are shown in Figure 20 [79]. During exposure to 100 ppm CO at 400°C, the sensor resistance decreased, which indicates that the SnO_2 thin films were n-type semiconductor sensor. From the difference between the sensor resistance in air (R_{air}) and the sensor resistance in CO (R_{CO}), we can predict the sensitivity of the sensors (sensitivity = R_{air} / R_{CO}). The sensitivity of graphene incorporated SnO_2 sensor was higher than the pristine SnO_2 sensor because the difference between R_{air} and R_{CO} increased compared to pristine SnO_2 sensor as shown in Figure 21(a) [79]. Sensing properties such as sensitivity, 90% response time, and 90% recovery time of nanostructure incorporated SnO_2 sensor were shown in Figure 21(b) [68]. The sensitivity, 90% response time, and 90% recovery time of the pristine SnO_2 sensor were 3.65, 10.04 s and 7.04 s, respectively. The sensitivity of the SnO_2 containing graphene sensor was increased to 6.84. Because of that SnO_2 thin films was oxidized in air and reduced in CO gas, it has lower R_{CO} and higher R_{air}. In case of sensor containing graphene, R_{CO} was decreased than pristine SnO_2 sensor. When the air flowed, SnO_2 thin film was oxidized by oxygen in air and the resistance was increased. Although graphene incorporated SnO_2 sensor was also oxidized, the difference between R_{CO} and R_{air} was increased than those of pristine SnO_2 sensor. The reason was increased surface area by graphene. Graphene have a large surface to volume ratio, the surface area of the film was increased upon graphene incorporation [54]. The larger surface area allows more gas molecules to be absorbed on the surface of the sensing film. By this reason, sensitivity of graphene incorporated SnO_2 sensor was increased despite low R_{air}.

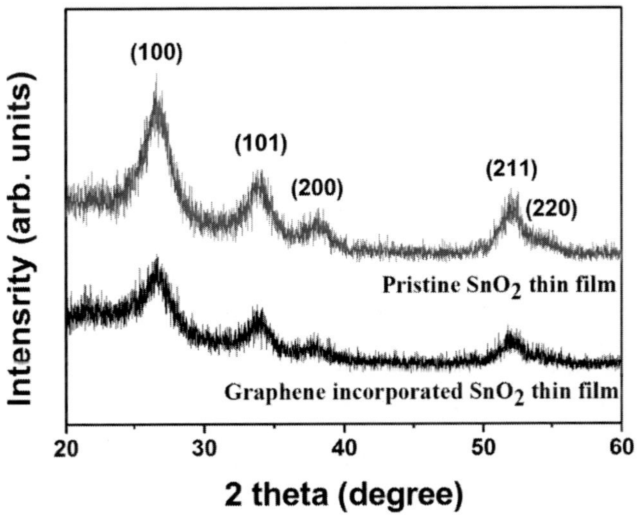

Figure 19. XRD patterns of SnO_2 hybrid films annealed at 500°C with 0.05 wt.% graphene.

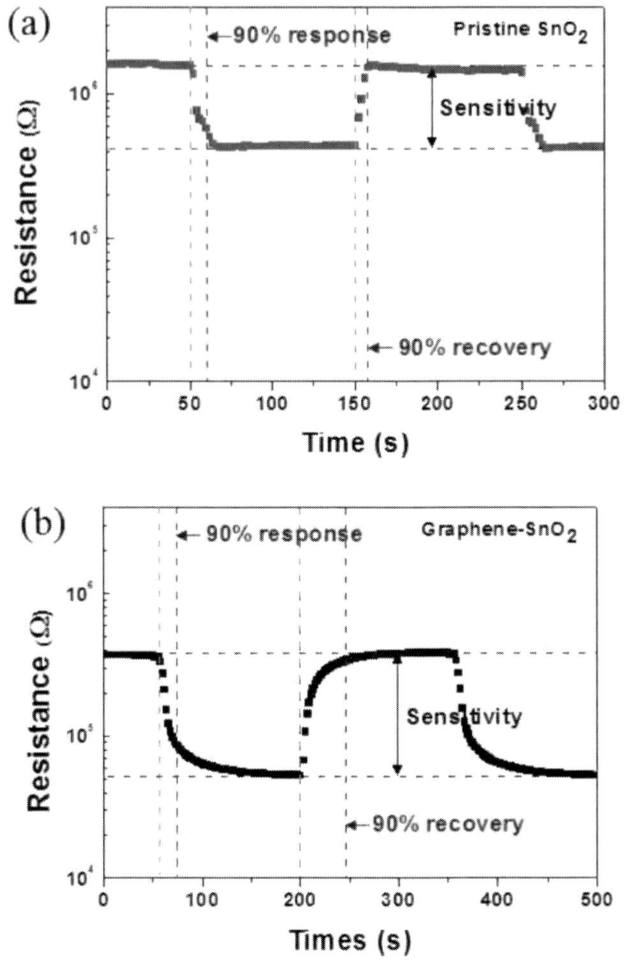

Figure 20. Transient response transient of (a) pristine SnO_2 and (b) graphene incorporated SnO_2 thin film sensor measured at 400°C with 100 ppm CO gas.

CONCLUSION

The ZnO and SnO_2 were chose as alterative materials of ITO considering its wide bandgap for transparency in visible light region. For applying to devices instead of ITO, the enhancement of electrical conductivity of the films is required. Graphene was incorporated into transparent conducting oxide films for increasing of electrical conductivity. And direct-patterning process was applied to this study for preventing damages on the physical and electrical property of functional films by using conventional etching process. The X-ray photoelectron spectroscopy, X-ray diffraction analysis, Hall effect analysis and synchrotron accelerator were used for interpreting structural and chemical bonding properties in conjunction with the electrical characteristics.

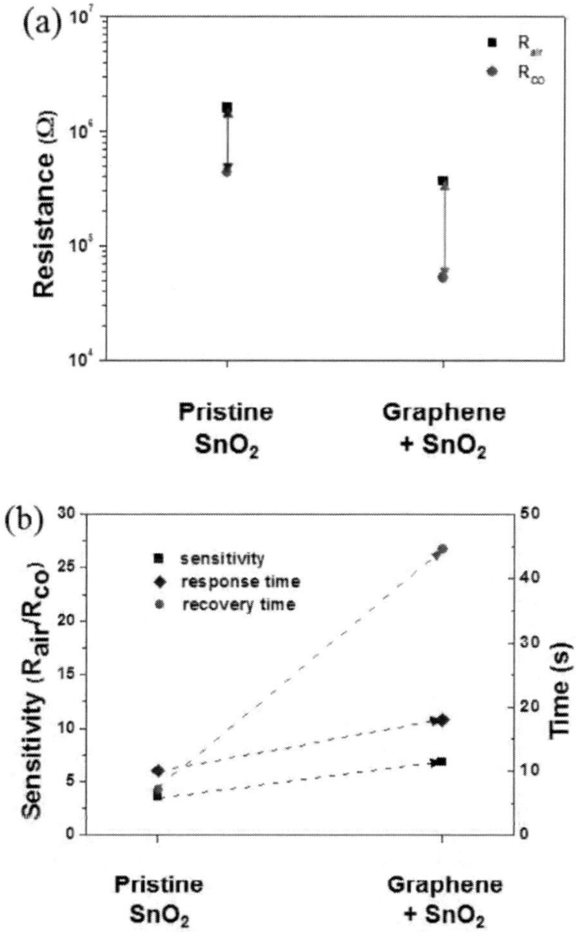

Figure 21. (a) Sensor resistance of nanostructure incorporated SnO_2 sensor in dry air or CO gas and (b) sensitivity (■), 90% response time (♦), and 90% recovery time (•) of nanostructure incorporated SnO_2 sensors measured at 400°C in dry air and 100 ppm CO gas.

Graphene was incorporated into transparent conducting oxide to enhance electrical conductivity because graphene has superior conductivity due to π-bonds on surface. The electrical resistivity of graphene incorporated transparent conducting oxide films was decreased due to mobility increasing by π-bonds on the surface of carbon nanostructure without significant degradation of crystallinity and transmittance. From the results by core level spectra analysis using XPS, it was confirmed that π-bonds at carbon nanostructure surface was maintained without any chemical bond between carbon nanostructure and transparent conducting oxide matrix.

During the study of incorporation graphene into transparent conducting oxide film, direct-patterning process was included to thin film fabrications by using photosensitive metal-organic precursor or photosensitive additive. The possibility of direct-patterning of graphene incorporated transparent conducting oxide film was examined by removing the area unexposed to UV light using a solvent after UV exposure. Direct-patterning of transparent conducting oxide film containing graphene with a micro-scaled line width was successfully performed without photoresist or dry etching.

Gas sensor application of graphene incorporated transparent conducting oxide film was studied. Gas reaction on surface is important factor in gas sensing performance. Due to a high electrical conductivity and surface area induced by graphene incorporation, the sensitivity of graphene incorporated oxide thin film sensors was enhanced relative to a pristine oxide thin film sensor.

REFERENCES

[1] Minami, T. (2005). Semicond. *Sci. Technol.*, *20*, S35.

[2] Lee, D., Bae, W. K., Park, I., Yoon, D. Y., Lee, S. & Lee, C. (2011). *Sol. Energ. Mat. Sol. C*, *95*, 365.

[3] Diniz, A. S. A. C. (2011). *Renew. Energ.*, *36*, 1153.

[4] Kang, D. W., Kuk, S. H., Ji, K. S., Lee, H. M. & Han, M. K. (2011). *Sol. Energ. Mat. Sol. C*, *95*, 138.

[5] Kim, H., Horowitz, J. S., Kim, W. H., Makinen, A. J., Kafafi, Z. H. & Chrisey, D. B. (2002). *Thin Solid Films*, *420*, 539.

[6] Minami, T., Nanto, H. & Takata, S. (1984). *Jpn. J. Appl. Phys.*, *23*, L280.

[7] Simakov, V. V., Yakusheva, O. V., Grebennikov, A. I. & Kisin, V. V. (2005). *Tech. Phys. Lett.*, *31*, 339.

[8] Carcia, P. F., McLean, R. S., Reilly, M. H. & Nunes Jr., G. (2003). *Appl. Phys. Lett.*, *82*, 1117.
[9] Kroger, F. A. (1964). Chemistry of Imperfect Crystals, North-Holland, Amsterdam, *Chaps.*, *7*, 16.
[10] Heiland, G., Mollow, E. & Stockmann, F. (1959). *SolidState Phys.*, *8*, 193.
[11] Ghandi, S. K., Field, R. J. & Shelly, J. R. (1980). *Appl. Phys. Lett.*, *37*, 449.
[12] Aranovich, J., Oritz, A. & Bube, R. H. (1979). *J. Vac. Sci, Technol.*, *16*, 994.
[13] Roth, A. P. & Williams, D. F. (1981). *J. Appl. Phys.*, *52*, 6685.
[14] Lehmann, H. W. & Widmer, R. (1974). Proc. 6th Int. Vacuum Congr., Kyoto, 1974, in *Jpn. J. Appl. Phys. Suppl.*, *2*, Part 1, 741.
[15] Lehmann, H. W., & Widmer, R. (1973). *J. Appl. Phys.*, *44*, 3868.
[16] Webb, J. B., Williams, D. F. & Buchanan, M. (1981). *Appl. Phys. Lett.*, *39*, 640.
[17] Look, D. C., Hemsky, J. W. & Sizelove, J. R. (1999). *Phys. Rev. Lett.*, *82*, 2552.
[18] Kato, H. (2002). *J. Cryst. Growth*, *237*, 538.
[19] Maudes, J. S. & Rodriguez, T. (1980). *Thin Solid Films*, *69*, 183.
[20] Grosse, P., Schmitte, F. J., Frank, G. & Köstlin, H. (1982). *Thin Solid Films*, *90*, 309.
[21] Shanthi, E., Dutta, V., Banerjee, A. & Chopra, K. L. (1980). *J. Appl. Phys.*, *51*, 6243.
[22] Advani, G. N., Jordan, A. G. & Kluge-Weiss, P. (1979). *Mater. Sci. Eng.*, *4*, 99.
[23] Sundaram, K. B. & Bhagavat, G. K. (1981). *Thin Solid Films*, *78*, 35.
[24] Heeq, M. & Portier, E. (1972).*Thin Solid Films*, *9*, 341.
[25] Leja, E., Pisarkiewicz, T. & Kotodziej, A. (1980). *Thin Solid Films*, *67*, 45.
[26] Manifaciet, J. C., Szepessy, L., Bresse, J. F. & Perotin, M. (1979). *Mater. Res. Bull.*, *14*, 109.
[27] Aboaf, J. A., Marcotte, V. C. & Chou, N. J. (1973). *J. Electrochem. Soc.*, *120*, 701.
[28] Raccanelli, A. & Maddalena, A. (1976). *J. Am. Ceram. Soc.*, *59*, 425.
[29] Spence, W. (1967). *J. Appl. Phys.*, *38*, 3767.
[30] Sundaram, K. B. & Bhagavat, G. K. (1981). *J. Phys. D*, *14*, 921.
[31] Inagaki, T., Najikama, J. & Nishimura, Y. (1969). *Fujitsu Sci. Tech. J.*, *5*, 235.

[32] Manifacier, J. C., de Murcia, M., Fillard, J. P. & Vicario, E. (1977). *Thin Solid Films*, *41*, 127.
[33] Miloslavskii, V. K. (1959). *Opt. Spectrosc.* (U.S.S.R.), *7*, 154.
[34] Manifacier, J. C., Fillard, J. P. & Bind, J. M. (1981). *Thin Solid Films*, *77*, 67.
[35] Kuznetsov, A. Ya. (1960). *Sov. Phys. Solid State*, *2*, 30.
[36] Gao, M. & Hill, R. H. (1998). *J. Mater. Res.*, *13*, 1379.
[37] Park, H. -H., Kim, W. S., Yang, J. K., Park, H. -H. & Hill, R. H. (2004). *Microelectron. Eng.*, *71*, 215.
[38] Park, H. -H., Park, H. -H. & Ross H. Hill. (2006). *Sensor. Actuat. A-Phys.*, *132*, 429.
[39] Kim, H., Wang, S. J., Park, H. H., Chang, H. J., eon, H. & Hill, R. H. (2007). *Thin Solid Films*, *516*, 198.
[40] Blair, S. L., Chu, C. W., Dammel, R. & Hill, R. H. (1997). *Proc. SPIE*, *3049*, 829.
[41] Andronic, L. S. & Hill, R. H. (2002). *J. Photochem. Photobiol. A: Chem.*, *152*, 259.
[42] H.-H. Park, H.-H. Park, Ross H. Hill. (2008). *Kor. J. Mater. Res.*, *18*, 2.
[43] Hwang, J. -S., Kim, W. S., Park, H. -H. & Kim, T. –S. (2003). *Journal of the Korean Ceramic Society*, *40*, 4, 385.
[44] Akhavan, O. (2011). *Carbon*, *49*, 11.
[45] Chastain, J. (Ed.). (1992). Handbook of X-ray Photoelectron Spectroscopy, Perkin-Elmer Corporation Physical Electronics Division, Minnesota, p. 40, p. 252.
[46] Yang, D., Velamakanni, A., Bozoklu, G., Park, S., Stoller, M., Piner, R. D., Stankovich, S., Jung, I., Field, D. A., Ventrice Jr., C. A. & Ruoff, R. S. (2009). *Carbon*, *47*, 145.
[47] Kundu, S., Wang, Y., Xia, W. & Muhler, M. (2008). *J. Phys. Chem. C*, *112*, 16869.
[48] Kotsis, K. & Staemmler, V. (2006). *Phys. Chem. Chem. Phys.*, *8*, 1490.
[49] Das, J., Pradhan, S. K., Sahu, D. R., Mishra, D. K., Sarangi, S. N., Nayak, B. B., Verma, S. & Roul, B. K. (2010). *Physica B*, *405*, 2492.
[50] Zie, J. & Varadan, V. K. (2005). *Mater. Chem. Phys.*, *91*, 274
[51] Huang, C. J., Li, C. H., Wang, H. L. & Lin, T. N. (2010). *Wrld. Acad. Sci. E.*, *65*, 266.
[52] Akhavan, O., Azimirad, R. & Safa, S. (2011). *Mater. Chem. Phys.*, *130*, 598.
[53] Zhang, H., Xu, P., Du, G., Chen, Z., Oh, K., Pan, D. & Jiao, Z. (2011). *Nano Res.*, *4*, 274.

[54] Mar, L. G., Timbrell, P. Y. & Lamb, R. N. (1993). *Thin Solid Films*, *223*, 341.

[55] McHale, A. E. (Ed.). (1996). Phase Equilibria Diagram; Volume XII Oxides, *The American Ceramic Society*, Ohio.

[56] Abdullah, H., Selmani, S., Norazia, M. N., Menon, P. S., Shaari, S. & Dee, C. F. (2011). *Sains Malays.*, *40*, 245.

[57] Joint Committee for Powder Diffraction Studies (JCPDS) No. 36-1451.

[58] Chen, J., Jang, C., Xiao, S., Ishigami, M. & Fuhrer, M. S. (2008). Nat. Nanotechnol., *3*, 206.

[59] Yan, Q., Wu, J., Zhou, G., Duan, W. & Gu, B. L. (2005). *Phys. Rev. B*, *72*, 155425.

[60] Bourlon, B., Miko, C., Forró, L., Glattli, D. C. & Bachtold, A. (2004). *Phys. Rev. Lett.*, *93*, 176806.

[61] Stankovich, S., Dikin, D. A., Dommett, G. H. B., Kohlhaas, K. M., Zimney, E. J., Stach, E. A., Piner, R. D., Nguyen, S. T. & Ruoff, R. S. (2006). *Nature*, *442*, 282.

[62] Nair, R. R., Blake, P., Grigorenko, A. N., Novoselov, K. S., Booth, T. J., Stauber, T., Peres, N. M. R. & Geim, A. K. (2008). *Science*, *320*, 1308.

[63] Marinopoulos, A. G., Wirtz, L., Marini, A., Olevano, V., Rubio, A. & Reining, L. (2004). *Appl. Phys. A*, *78*, 1157.

[64] Grodecki, K., Drabińska, A., Bożek, R., Wysmołeka, A., Korona, K. P., Strupiński, W., Borysiuk, J., Stępniewski, R. & Baranowski, J. M. (2009). *Acta. Phys. Pol. A*, *116*, 835.

[65] Tang, Y. & Gou, J. (2010). *Mater. Lett.*, *64*, 2513.

[66] Park, N., Kim, B. K., Lee, J. O. & Kim, J. J. (2005). *Appl. Phys. Lett.*, *95*, 243105.

[67] Zengin, H., Zhou, W., Jin, J., Czerw, R., Smith, D. W., Echegoyen, L., Carroll, D. L., Foulger, S. H. & Ballato, J. (2002). *Adv. Mater.*, *14*, 1480.

[68] Levin, E. M., Robbins, C. R. & McMurdie, H. F. (1981). *Phase diagrams for ceramists*, Volume *4* (The American ceramic so-ciety, Columbus), p.10.

[69] Joint Committee for Powder Diffraction Studies (JCPDS) No. 70-4177.

[70] V. C. Tung, L. M. Chen, M. J. Allen, J. K. Wassei, K. Nel-son, R. B. Kaner, and Y. Yang, (2009). *Nano Lett*, 9, 1949.

[71] Kim, H., Yun, M. K. & Park, H. H. (2011). *Phys. Status Solidi A*, *208*, 1869.

[72] Robertson, J. & O'Reilly, E. P. (1987). *Phys. Rev. B*, *35*, 2946.

[73] Robertson, J. (2002). *Mat. Sci. Eng. R*, *37*, 129.

[74] Fortunato, E., Raniero, L., Silva, L., Gonçalves, A., Pimentel, A., Barquinha, P., Águas, H., Pereira, L., Gonçalves, G., Ferreira, I., Elangovan, E. & Martins, R. (2008). *Sol. Energ. Mat. Sol. C.*, *92*, 1605.

[75] Hudson, J. B. (1992). Surface science: an introduction (Butterworth-Heinemann, Stoneham), p.38.

[76] Neamen, D. A. (2003). Semiconductor physics and devices: basic principles (McGraw-Hill, New York), p.103.

[77] Tameev, A. R., Jiménez, L. L., Pereshivko, L. Y., Rychwalski, R. W. & Vannikov, A. V. (2007). *J. Phys.: Conf. Ser.*, *61*, 1152.

[78] Lee, J. -S., Ha, T. -J., Hong, M. –H. & Park, H. -H. (2013). *Thin Solid Films*, *529*, 98.

[79] Kim, H. (2012). *a doctor´s thesis*, Yonsei University, December.

In: Graphene
Editor: Bruce T. Edwards
ISBN: 978-1-62948-795-3

Chapter 6

Graphene and Related Nanomaterials for Environmental Remediation

Shamik Chowdhury and Rajasekhar Balasubramanian
Department of Civil and Environmental Engineering,
National University of Singapore, Singapore, Republic of Singapore

Abstract

It has been widely acknowledged that finding solutions to environmental problems that we are facing now has become a top priority. Environmental problems are highly complex and often vary in scale ranging from local to global dimensions. The advent of graphene revolution in recent years has provided immense scope and opportunities for environmental remediation. Graphene, and its related materials such as graphene oxide, reduced graphene oxide and their nanocomposites can all function as adsorbents and catalysts for the effective removal and degradation of recalcitrant organic pollutants (e.g., dyes, phenolic compounds, polycyclic aromatic hydrocarbons), toxic inorganic contaminants (e.g., heavy metals) and harmful biological substances (e.g., bacteria, microcystins), present invariably in the aquatic environment. Graphene-based materials are also popular in decontaminating industrial waste gaseous streams containing various air pollutants such as NH_3, H_2-S, SO_x, NO_x, etc. This chapter aims at bringing together the current knowledge available in the development and use of graphene and its related nanomaterials for environmental clean-up. This synthesis of advanced knowledge in graphene, an emerging field of interest, will help in fuelling further innovations in the development of novel graphene-based nanomaterials for environmental pollution control and abatement, a topic of current global interest.

Keywords: Graphene, pollution, adsorption, photocatalysis, nanomaterials, environmental remediation

1. INTRODUCTION

Environmental pollution is considered to be of global concern. The rapid pace of industrialization, exponential population growth and unplanned urbanization have resulted in environmental pollution with toxic materials (heavy metals, synthetic dyes, persistent organic pollutants, etc.), greenhouse gases (carbon dioxide, methane, nitrous oxide) and other poisonous gas fumes and smoke (volatile organic compounds and hydrofluorocarbons), threatening humans and ecosystems with serious health risks. Preventing environmental pollution and protecting the environment is therefore a top priority. Nanomaterials show a better performance in environmental remediation than other conventional materials because of their high surface area (surface-to-volume ratio) and their associated high reactivity (Khin et al., 2012). In particular, carbon nanomaterials such as carbon nanotubes (CNTs), carbon nanofibers (CNFs), carbon nanospheres (CNSs), nanodiamond and fullerene have gained a great deal of attention for treatment of pollutants in water and air (Wang et al., 2013a).

Recently, graphene as a newly discovered carbon nanomaterial has emerged as a "wonder material" with enormous potential for environmental pollution control and abatement. Graphene is basically a flat single layer of sp^2 hybridized carbon atoms, densely packed into an ordered two-dimensional honeycomb network (Figure 1) (Ivanovskii, 2012). This one-atom thick allotrope of elemental carbon can be viewed as the basic structural unit of other carbon allotropes. It can be wrapped into zero-dimensional fullerenes, rolled into one-dimensional CNTs, or can be stacked into three-dimensional graphite (Figure 2) (Brownson et al., 2011). A unit hexagonal cell of graphene comprises two equivalent sub-lattices of carbon atoms, joined together by σ bonds with a carbon–carbon bond length of 0.142 nm (Figure 1) (Avouris and Dimitrakopoulos, 2012). Each carbon atom in the lattice has a π orbital that contributes to a delocalized network of electrons, making graphene sufficiently stable compared to other nanosystems (Zhu et al., 2010). Theoretical and experimental studies have proved that graphene offers a unique combination of high three-dimensional aspect ratio and large specific surface area, superior mechanical stiffness and flexibility, remarkable optical transmittance, exceptionally high electronic and thermal conductivities, impermeability to gases, as well as many other supreme properties (Novoselov et al., 2012). Because of these fascinating properties, graphene has sparked enormous scientific interest in realizing its many exciting and revolutionary applications. The wide range of graphene's applications includes: nanoelectronics (Ruoff, 2008), structural composites (Stankovich et al., 2006), conducting polymers (Stankovich et al., 2006), battery electrodes (Paek et al., 2009; Su et al., 2010), supercapacitors (Dikin et al., 2007), transport barriers (Bunch et al., 2008; Compton et al., 2010), printable inks (Wang et al., 2010), antibacterial papers (Dikin et al., 2007), and biomedical technologies (Zhang et al., 2010a; Ryoo et al., 2010). More recently, the multifarious applications of graphene have encouraged not only the development of substrate-bound extended graphene monolayers, but also related materials such as graphene oxide (GO), reduced graphene oxide (rGO), and few-layered graphene oxide (FGO) (Sanchez et al., 2012). On account of their intriguing physical and chemical properties, there has been a steadily growing interest in the use of these materials to degrade and scavenge pollutants in water and air, reflected by an enormous increase in the number of research articles in the recent literature. Graphene and its derivatives have proven to be excellent adsorbents and catalysts due to their large specific surface areas and high reactivities. This chapter focuses on reviewing the current knowledge on the use of this novel and versatile family of nanomaterials in pollution management with an emphasis on gas adsorption and water remediation. The latest advances in experimental studies and relevant published data on decontamination of wastewater and waste-gas have been summarized and presented. The current challenges in the field have also been critically examined to help identify future research trends and opportunities.

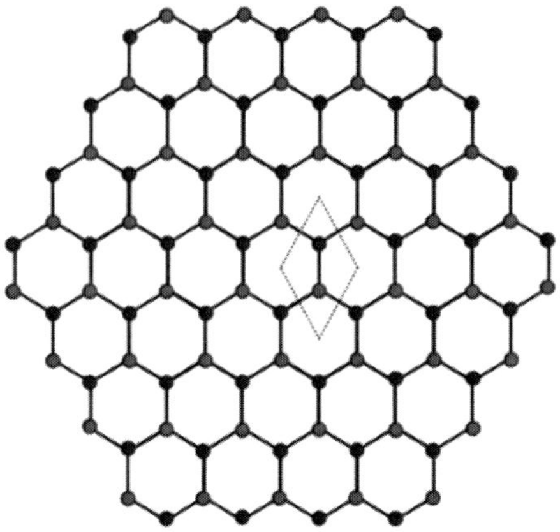

Figure 1. Representation of the honeycomb lattice of graphene and its unit cell (indicated by the dashed line). The unit cell contains two atoms, each one belonging to a different sub-lattice.

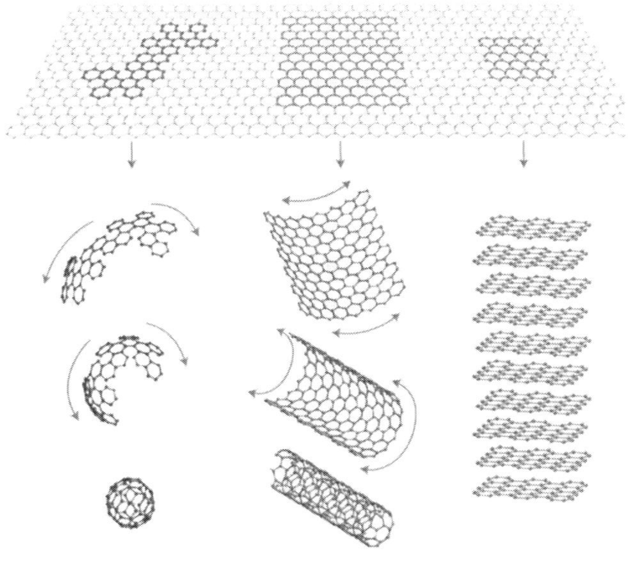

Figure 2. Graphene as a basic building block for graphitic materials of all other dimensionalities. It can be wrapped up into 0-dimensional fullerenes, rolled into 1-dimensional nanotubes or stacked into 3-dimensional graphite. Reprinted from Geim and Novoselov (2007), Copyright 2007, with permission from Macmillan Publishers Ltd.

2. Water Treatment and Remediation

2.1. Water Remediation by Adsorption

In both developing and industrialized nations, the surge of industrial, agricultural and domestic activities has inevitably resulted in an increased flux of toxic pollutants into the surrounding water bodies (Wang and Chen, 2009). Freshwater can be contaminated with a variety of pollutants ranging from heavy metals, dyes, phenolic compounds, pesticides, herbicides to emerging micropollutants such as endocrine disrupting compounds (EDCs), pharmaceuticals and personal care products (PPCPs) and nitrosamines (Shannon et al., 2008; Lapworth et al., 2012). These pollutants can have bio-accumulative, persistent, carcinogenic, mutagenic and detrimental effects on the survival of aquatic organisms, flora, fauna as well as human health (Corcoran et al., 2010).

With the enforcement of stringent rules and regulations concerning the uncontrolled discharge of toxic pollutants, there has been a recent flurry of activity in water treatment research leading to the development of a wide array of wastewater treatment techniques. Of all the technologies that have been proposed, adsorption is widely acknowledged as the most promising and efficient method because of its low capital investment, simplicity of design, ease of operation, insensitivity to toxic substances and complete removal of pollutants even from dilute solutions (Dabrowski, 2001; Crini, 2005; Crini and Badot, 2008; Foo and Hameed, 2010; Bhatnagar and Sillanpaa, 2010). Adsorption treatment also does not result in any harmful substances and produces a high quality treated effluent. Activated carbon, a crude form of graphite, is the adsorbent of choice due to its highly porous structure and large surface area (Gupta and Suhas, 2009; Malaviya and Singh, 2011). It is extensively used for the removal of different types of pollutants from drinking water such as metal ions, dyes, phenols, pesticides, chlorinated hydrocarbons, humic substances, detergents, polychlorinated biphenyls (PCBs), polycyclic aromatic hydrocarbons (PAHs) and even micropollutants (Gupta et al., 2009; Delgado et al., 2012; Djilani et al., 2012). However, the widespread use of activated carbon is restricted due to economic considerations (Gupta et al., 2009). Attempts have, therefore, been made by many researchers to find inexpensive alternatives to activated carbon. Most research undertaken for that purpose have focused on the use of waste/by-products from industries and agricultural operations, natural materials, or microbial and non-microbial biomass (Aksu, 2005; Kurniawan et al., 2006; Crini, 2006; Ahmaruzzaman,

2008; Demirbas, 2008). However, these low-cost adsorbents are largely criticized for their low adsorption capacities and potential disposal problems and have thus not been applied at an industrial scale (Crini, 2006). Therefore, the exploration of new promising adsorbents is still in progress.

Recently graphene and its related materials are being intensively studied as a new type of adsorbents for removal of toxic pollutants from aquatic systems. This extraordinary interest stems from their unique morphology, nanosized scale and novel physico-chemical properties (Wang et al., 2013a). Their high specific surface areas are comparable to those of activated carbon. They also have far more well-defined and uniform structure than activated carbon and most other adsorbent materials reported till date. In addition, these nanomaterials are not only capable of sequestering contaminants with varying molecular sizes, but also have considerably high adsorption capacities and can be easily regenerated for repeated use.

The potential application of graphene materials as adsorbents strongly depends on their homogeneous dispersion in the liquid phase as well as their ability to remove different types of contaminants. However, graphene as a bulk material has the tendency to agglomerate and restack to form graphite during liquid processing (Cheng et al., 2012). On the other hand, GO has a weak binding affinity for anionic compounds due to its strong electrostatic repulsion between them. Additionally, both graphene and GO cannot be easily collected and separated from treated water, leading to serious recontamination.

Chemical functionalization of graphene materials is an effective and practical approach which can actually facilitate the dispersion and stabilize graphene to prevent agglomeration (Kuilla et al., 2010). It also can improve their processability and enhance their interaction with different types of substances. The functional groups attached can be nanosized metal oxides (NMOs) such as nanosized TiO_2, ZnO, MnO_2, SiO_2, Fe_3O_4, and $CoFe_2O_4$, or organic polymers such as chitosan, polystyrene, polyaniline, polyurethane and polycaprolactone (Huang et al., 2012). The different types of polymers or nanoparticles can be directly decorated on the graphenic sheets, and no molecular linkers are needed to bridge the polymers/nanoparticles and the graphenic sheets which may prevent additional trap states along the sheets (Singh et al., 2011). The resulting nanocomposite is not merely the sum of the individual components, but instead a new material with new functionalities and properties (Wu et al., 2012). The NMO/polymer anchored on graphene two-dimensional structures not only prevents the agglomeration and restacking, but also increases the available surface area of the graphene sheet alone, leading to high adsorption activity. The incorporated material also provides high selectivity and strong binding of the desired contaminant depending on its structure, size and crystallinity. In contrast, the graphenic materials provide chemical functionality and compatibility to allow easy processing of the deposited NMO/polymer in the composite. The ultimate goal is to maximize the practical use of the combined advantages of both the components as active materials for improving the adsorption performance and potential.

2.1.1. Adsorption of Inorganic Pollutants

At the forefront of graphene revolution, numerous investigations have been undertaken to evaluate the metal removal capability of graphene materials from water systems (Table 1). Leng et al., (2012) explored the possibility of using graphene as an adsorbent for the removal of Sb(III) from aqueous solution. The adsorption capacity decreased with increasing metal ion concentrations, whereas it increased with increasing temperature. A sharp increase in the metal removal efficiency was observed with increase in pH beyond 3.8. A maximum removal of about 99.5% was observed at pH>11. Adsorption phenomena appeared to follow the Freundlich isotherm better than the Langmuir isotherm. Under the optimized conditions, the adsorption capacity of graphene for Sb(III) was found to be 10.919 mg g^{-1}. The adsorption kinetic data best fitted the pseudo-second-order rate expression. The metal loaded graphene could be easily regenerated using 0.1 mol L^{-1} EDTA as a desorbing agent. Even after five consecutive cycles of adsorption–desorption, a removal efficiency of 60% could be attained.

Adsorption of U(VI) from aqueous solution onto GO was investigated by Li et al., (2012a). The adsorption process was independent of ionic strength and a contact time of 1 h was found sufficient to achieve equilibrium conditions. The adsorption isotherm data could be well fitted to the Langmuir model. The maximum U(VI) adsorption capacity of GO was determined to be 299 mg g^{-1} which was significantly higher than that of many other reported adsorbents. Thermodynamic studies indicated endothermic and spontaneous nature of adsorption of U(VI) by GO.

Wu et al., (2013b) tested the adsorption capacity of GO for the removal of Cu(II) from aqueous solution. The optimum conditions for Cu(II) removal by GO in a batch system were: solution pH 5.3, adsorbent dosage of 1 mg mL^{-1} and an equilibrium contact time of 150 min. The Cu(II) equilibrium adsorption data fitted well to the Freundlich isotherm model. Based on the Langmuir model constants, GO exhibited a very high adsorption capacity of 117.5 mg g^{-1} for Cu(II). The adsorption of Cu(II) onto GO was mainly attributed to surface complexation, ion exchange, and electrostatic attraction. Desorption experiments demonstrated that more than 74% of the adsorbed Cu(II) could be desorbed at pH<1 using HCl as the desorbing agent. GO retained more than 90% of its initial Cu(II) adsorption capacity even after ten adsorption-desorption cycles, implying its repeated use in the removal of heavy metals from industrial wastewater.

Wang et al., (2013b) demonstrated the ability of GO to adsorb Zn(II) ions from its aqueous solution. The adsorption of Zn(II) on GO was strongly dependent on pH but was weakly affected by foreign ions and ionic strength. The Zn(II) adsorption process followed the pseudo-second-order kinetics. Equilibrium adsorption data obtained at different temperatures were well described by the Langmuir isotherm with a maximum monolayer adsorption capacity of 246 mg g^{-1} at 293 K. The thermodynamic parameters obtained from the temperature dependent adsorption isotherms indicated that the adsorption reaction was a spontaneous and exothermic process.

The feasibility of using GO for removal of divalent metal ions of Cu, Zn, Cd and Pb was recently investigated by Sitko et al., (2013). The adsorption affinity was found to follow the sequence Pb(II) > Cd(II) > Zn(II) > Cu(II) in single-metal systems, but followed the order Pb(II) > Cu(II) >> Cd(II) > Zn(II) in binary-metal systems. Adsorption isotherm and kinetic studies showed that adsorption of metal ions onto GO was monolayer coverage and controlled by chemisorption involving strong surface complexation of metal ions with the oxygen-containing groups on the surface of GO.

GO aerogels prepared by a unidirectional freeze-drying method has also been studied for the removal of metal ions (Cu^{2+}) from aqueous solution by Mi et al., (2012). GO aerogel showed a very fast adsorption rate due to its extended layered structure with interconnected pores (Figure 3). Maximum amount of Cu(II) was removed within the first 15 min of metal-adsorbent contact. Cu(II) adsorption kinetic data could be best modeled by the pseudo-second-order rate equation indicating that adsorption of Cu(II) onto GO aerogel was controlled by chemical adsorption involving valence forces through sharing or exchange of electrons between the adsorbate and the adsorbent. The adsorption equilibrium data were better described by the Langmuir model than the Freundlich model. The maximum Cu(II) adsorption capacity increased from 17.73 mg g^{-1} at 283 K to 29.59 mg g^{-1} at 313 K, indicating an endothermic nature of the adsorption process.

Table 1. Reported results of batch adsorption studies on the removal of heavy metals from water by graphene and its related materials

Adsorbent	Adsorbate	Conc.	pH	Temp. (K)	Contact time (h)	Adsorption capacity	Reference
Graphene	Sb(III)	1-10 mg L^{-1}	11.0	303	4	10.919 mg g^{-1}	Leng et al., 2012
CTAB modified Graphene	Cr(VI)	20-100 mg L^{-1}	2.0	293	1	21.57 mg g^{-1}	Wu et al., 2013a
Functionalized Graphene (GNS^{PF6})	Pb(II)	—	5.1	—	4	406.4 mg g^{-1}	Deng et al., 2010
Functionalized Graphene (GNS^{PF6})	Cd(II)	—	6.2	—	4	73.42 mg g^{-1}	Deng et al., 2010
Functionalized Graphene (GNS^{C8P})	Pb(II)	—	5.1	—	4	74.18 mg g^{-1}	Deng et al., 2010
Functionalized Graphene (GNS^{C8P})	Cd(II)	—	6.2	—	4	30.05 mg g^{-1}	Deng et al., 2010
GO	U(VI)	—	4.0	Room temp.	4	299 mg g^{-1}	Li et al., 2012a
GO	Cu(II)	25-250 mg L^{-1}	5.3	—	2.5	117.5 mg g^{-1}	Wu et al., 2013b
GO	Zn(II)	10-100 mg L^{-1}	7.0 ± 0.1	293	—	246 mg g^{-1}	Wang et al., 2013b
GO	Cu(II)	—	5.0	298	2	294 mg g^{-1}	Sitko et al., 2013
GO	Zn(II)	—	5.0	298	2	345 mg g^{-1}	Sitko et al., 2013
GO	Cd(II)	—	5.0	298	2	530 mg g^{-1}	Sitko et al., 2013
GO	Pb(II)	—	5.0	298	2	1119 mg g^{-1}	Sitko et al., 2013
GO aerogel	Cu(II)	50-75 mg L^{-1}	6.3	313	0.5	29.59 mg g^{-1}	Mi et al., 2012
rGO	U(VI)	—	4.0	Room temp.	4	47 mg g^{-1}	Li et al., 2012a
FGO	Pb(II)	—	6.0	333	24	1850 mg g-1	Zhao et al., 2011a

Table 1. (Continued)

Adsorbent	Adsorbate	Conc.	pH	Temp. (K)	Contact time (h)	Adsorption capacity	Reference
FGO	Cd(II)	—	6.0 ± 0.1	333	—	167.5 mg g^{-1}	Zhao et al., 2011b
FGO	Co(II)	—	6.0 ± 0.1	333	—	79.8 mg g^{-1}	Zhao et al., 2011b
Iron nanoparticle decorated Graphene	Cr(VI)	25-125 mg L^{-1}	4.25	293	4	162 mg g^{-1}	Jabeen et al., 2011
Graphene/δ-MnO_2	Ni(II)	10-100 mg L^{-1}	—	318	3	66.01 mg g^{-1}	Ren et al., 2011
Graphene/δ-MnO_2	Cu(II)	—	6.0	298 ± 2	2	1637.9 µmol g^{-1}	Ren et al., 2012
Graphene/δ-MnO_2	Pb(II)	—	6.0	298 ± 2	2	793.65 µmol g^{-1}	Ren et al., 2012
Graphene/Fe@Fe_2O_3@ Si-S-O	Cr(VI)	1 g L^{-1}	7.0	—	—	1.03 mg g^{-1}	Zhu et al., 2012
SiO_2/Graphene	Pb(II)	20 mg L^{-1}	6.0	298	1	113.6 mg g^{-1}	Hao et al., 2012
EDTA/GO	Pb(II)	5-300 mg L^{-1}	6.8	298 ± 2	24	525 mg g^{-1}	Madadrang et al., 2012
Magnetite-rGO	As(V)	3-7 mg L^{-1}	7.0	293	2	13.10 mg g^{-1}	Chandra et al., 2010
Magnetite-rGO	As(III)	3-7 mg L^{-1}	7.0	293	2	10.20 mg g^{-1}	Chandra et al., 2010
GO/Ferric hydroxide	As(V)	0.5-20 mg L^{-1}	4.0-9.0	Room temp.	24	23.78 mg g^{-1}	Zhang et al., 2010b
GO/Chitosan	Pb(II)	50 mg L^{-1}	—	Room temp.	—	99 mg g^{-1}	He et al., 2011
Polypyrrole-rGO	Hg(II)	50-250 mg L^{-1}	3.0	293	3	979.54 mg g^{-1}	Chandra and Kim, 2011
GO- Iron oxide	Pb(II)	10-15 mg L^{-1}	6.5 ± 0.1	303	48	588.24 mg g^{-1}	Yang et al., 2012
Polypyrrole/GO	Cr(VI)	—	—	—	—	9.56 mmol g^{-1}	Li et al., 2012b

Adsorbent	Adsorbate	Conc.	pH	Temp. (K)	Contact time (h)	Adsorption capacity	Reference
GO-TiO_2 (hydrothermal treatment at 373 K for 6h)	Zn(II)	—	5.6	—	—	44.8 ± 3.4 mg g^{-1}	Lee and Yang, 2012
GO-TiO_2 (hydrothermal treatment at 373 K for 6h)	Cd(II)	—	5.6	—	—	65.1 ± 4.4 mg g^{-1}	Lee and Yang, 2012
GO-TiO_2 (hydrothermal treatment at 373 K for 6h)	Pb(II)	—	5.6	—	—	45.0 ± 3.8 mg g^{-1}	Lee and Yang, 2012
GO-TiO_2 (hydrothermal treatment at 373 K for 12h)	Zn(II)	—	5.6	—	—	88.9 ± 3.3 mg g^{-1}	Lee and Yang, 2012
GO-TiO_2 (hydrothermal treatment at 373 K for 12h)	Cd(II)	—	5.6	—	—	72.8 ± 1.6 mg g^{-1}	Lee and Yang, 2012
GO-TiO_2 (hydrothermal treatment at 373 K for 12h)	Pb(II)	—	5.6	—	—	65.6 ± 2.7 mg g^{-1}	Lee and Yang, 2012
GO@Sepiolite	U(VI)	10-50 mg L^{-1}	5.0	298	—	161.3 mg g^{-1}	Cheng et al., 2013
Magnetic chitosan/GO	Pb(II)	—	5.0	303 ± 0.2	1	76.94 mg g-1	Fan et al., 2013a
GO- GO-ZrO(OH)2	As(III)	2-80 mg L-1	7.0 ± 0.2	298.5 ± 0.2	0.25	95.15 mg g-1	Luo et al., 2013
GO- GO-ZrO(OH)2	As(V)	2-80 mg L-1	7.0 ± 0.2	298.5 ± 0.2	0.25	84.89 mg g-1	Luo et al., 2013
Fe3O4/GO	U(VI)	2.25 ×10-5 — 2.24 ×10-4 mol L-1	5.5± 0.1	293	24	69.49 mg g-1	Zong et al., 2013
GO-FeOOH	As(V)	—	7.0	298	—	73.42 mg g-1	Peng et al., 2013
Magnetic cyclodextrin–Chitosan/GO	Cr(VI)	50 mg L-1	3.0	323	—	67.66 mg g-1	Li et al., 2013a

Table 1. (Continued)

Adsorbent	Adsorbate	Conc.	pH	Temp. (K)	Contact time (h)	Adsorption capacity	Reference
Calcium alginate/GO	Cu(II)	—	—	Room temp.	1.5	60.24 mg g^{-1}	Algothmi et al., 2013
rGO-MnO_2	Hg(II)	1 mg L^{-1}	—	303 ± 2	—	9.50 mg g^{-1}	Sreeprasad et al., 2011
rGO-Ag	Hg(II)	1 mg L^{-1}	—	303 ± 2	—	9.53 mg g^{-1}	Sreeprasad et al., 2011
rGO- Iron oxide	Pb(II)	10-15 mg L^{-1}	6.5 ± 0.1	303	48	454.55 mg g^{-1}	Yang et al., 2012
rGO-Fe(0)	As(III)	2-6 mg L^{-1}	7.0	298	1	37.3 mg g^{-1}	Bhunia et al., 2012
rGO- Fe_3O_4	As(III)	2-6 mg L^{-1}	7.0	298	1	21.2 mg g^{-1}	Bhunia et al., 2012
rGO-Fe(0)/ Fe_3O_4	As(III)	2-6 mg L^{-1}	7.0	298	1	44.4 mg g^{-1}	Bhunia et al., 2012
rGO-Fe(0)/ Fe_3O_4	Cr(VI)	2-6 mg L^{-1}	7.0	298	1	31.1 mg g^{-1}	Bhunia et al., 2012
rGO-Fe(0)/ Fe_3O_4	Hg(II)	2-6 mg L^{-1}	7.0	298	1	22.0 mg g^{-1}	Bhunia et al., 2012
rGO-Fe(0)/ Fe_3O_4	Pb(II)	2-6 mg L^{-1}	7.0	298	1	19.7 mg g^{-1}	Bhunia et al., 2012
rGO-Fe(0)/ Fe_3O_4	Cd(II)	2-6 mg L^{-1}	7.0	298	1	1.91 mg g^{-1}	Bhunia et al., 2012

Figure 3. GO aqueous suspension (a); optical image of GO aerogel (b); low magnitude SEM image of GO aerogel (c); high magnitude SEM images of the GO aerogel, cross (d) and lateral section (e). Reprinted from Mi et al., (2012), Copyright 2012, with permission from Elsevier.[1]

Zhao et al., (2011a) reported that FGO nanosheets can be used to remove Pb(II) ions from aqueous solutions. In their study, 2- or 3-layered GO nanosheets were synthesized from graphite by using the Hummer's method. The prepared FGO nanosheets had numerous oxygen-containing functional groups on their surface and high dispersion properties in aqueous solution. The percent Pb(II) removal increased with increase in pH from 1.0 to 8.0, and further increase in pH reduced the metal adsorption efficiency. The experimental equilibrium data were found to conform to the Langmuir isotherm model, showing maximum monolayer adsorption capacity of about 1850 mg g^{-1} at 333 K. The high Pb(II) adsorption capacity of FGO was ascribed to the strong surface complexation between Pb(II) ions and the

---

[1] SEM: Scanning Electron Microscopy.

abundant oxygen-containing groups on the surfaces of FGO. Thermodynamic parameters were also calculated, and the adsorption was suggested to be spontaneous and endothermic. In a separate study by the same research group (Zhao et al., 2011b), FGO nanosheets were used as adsorbents for the removal of Cd(II) and Co(II). Adsorption of Cd(II) and Co(II) onto FGO nanosheets was strongly dependent on pH and weakly dependent on ionic strength. The presence of humic acid at pH<8 reduced the Cd(II) as well as Co(II) adsorption potential of FGO nanosheets. The Langmuir monolayer adsorption capacity was calculated to be around 167.5 and 79.8 mg g^{-1} for Cd(II) and Co(II), respectively at pH 6.0 ± 0.1 and temperature of 333 K. A thermodynamic evaluation indicated spontaneous and endothermic nature of adsorption of Cd(II) and Co(II) onto FGO nanosheets.

Several protocols to enhance the metal adsorption capacity of graphene materials have also been developed. Deng et al., (2010) reported an investigation of chemical functionalization of graphene to improve its adsorption efficiency for heavy metals. They adopted a mild, one-step electrochemical approach for the preparation of functionalized graphene sheets with the assistance of an ionic liquid and water. In a typical experiment, two high-purity graphite rods were inserted in parallel as electrodes into an ionic liquid/water solution under a static voltage of 15 V. 1-octyl-3-methylimidazolium hexafluorophosphate and potassium hexafluorophosphate were used as the ionic liquids; functionalized graphene sheets thus obtained were named GNS^{CP8} and GNS^{PF6}, respectively. Deng et al., (2010) then investigated the adsorption of two heavy metals, Pb(II) and Cd(II), onto GNS^{CP8} and GNS^{PF6}. GNS^{CP8} showed an adsorption efficiency of 34.38% and 22.75% for Pb(II) and Cd(II), at pH 5.80 and pH 6.30, respectively. GNS^{PF6}, on the other hand, exhibited a relatively high removal percentage of 89.89 % and 53.60 % for Pb(II) and Cd(II), at pH 5.80 and pH 6.10, respectively. The batch adsorption equilibrium data showed excellent fit to both the Langmuir and Freundlich isotherms. The pseudo-second-order model provided a better correlation in contrast to the pseudo-first-order model for the adsorption kinetic data of Pb(II)/Cd(II) onto GNS^{CP8} and GNS^{PF6}. Due to the better adsorption performance of GNS^{PF6} than GNS^{CP8}, regeneration possibilities of GNS^{PF6} was also considered. The regeneration experiments showed that GNS^{PF6} could be reused for five cycles without any significant loss in its initial adsorption capacity.

Madadrang et al. (2012) enhanced the adsorption capacity of GO by introducing chelating groups to its surface through a silanization reaction between N-(trimethoxysilylpropyl) ethylenediamine triacetic acid (EDTA-silane) and hydroxyl groups of GO. EDTA/GO proved to be an ideal adsorbent for removal of Pb(II) (Figure 4). The adsorption process was found to be pseudo-second-order, obeying the Langmuir adsorption isotherm. EDTA/GO exhibited a maximum Pb(II) adsorption capacity of 525 mg g^{-1} which was significantly higher than that of pristine GO (367 mg g^{-1}).

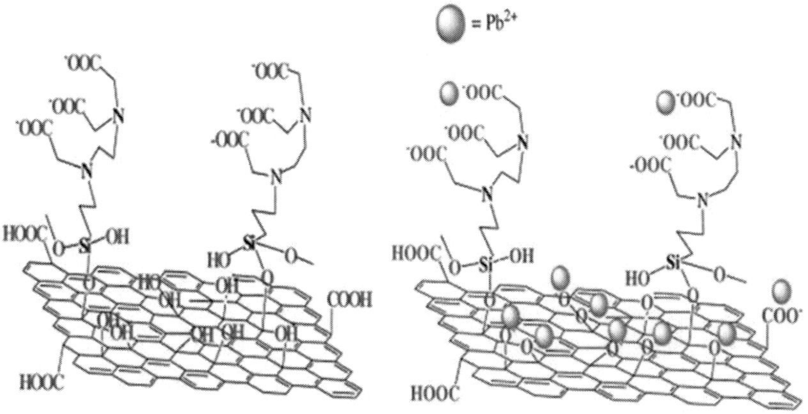

Figure 4. Chemical structure of EDTA-GO (left) and its interaction with Pb(II) ions (right). Reprinted from Madadrang et al. (2012), Copyright 2012, with permission from American Chemical Society.

A study to enhance the metal uptake capacity of graphene has recently been described by Wu et al., (2013a). They used cetyltrimethylammonium bromide (CTAB) to modify graphene which was then used as adsorbent to remove Cr(VI) from aqueous solutions. A removal efficiency of 98.2 % was obtained when 0.4 g of CTAB-modified graphene was used to adsorb hexavalent chromium from a test solution containing 50 mg L^{-1} of Cr(VI). Cr(VI) removal was also found to be pH dependent, reaching a maximum at pH 2.0. The adsorption process followed pseudo-second-order kinetics and equilibrium was attained in 1 h. The maximum adsorption capacity, calculated using the Langmuir model, was 21.57 mg g^{-1} at 293 K. The introduction of CTAB groups to the graphene surface significantly increased the adsorption capacity of graphene for heavy metals. A thermodynamic assessment suggested that adsorption of Cr(VI) by CTAB-modified graphene was a spontaneous and exothermic process.

A variety of nanocomposites of graphene materials (NGMs) with significant adsorption potential have also been developed for removal of heavy metal ions from aqueous environment. Jabeen et al., (2011) synthesized graphene sheets decorated with zero valent iron nanoparticles (G-nZVI) and used it for removal of Cr(VI) from aqueous solutions. The adsorption process was found to be pH dependent, showing maximum equilibrium adsorption at pH 2.0—3.0. Cr(VI) removal efficiency increased with increase in temperature from 283 to 323 K. Adsorption of Cr(VI) by G-nZVI was best described by the pseudo-second-order kinetic model. As much as 162 mg g^{-1} of Cr(VI) could be adsorbed as determined by the Langmuir isotherm model.

Zhang et al., (2010b) developed GO/ferric hydroxide composites showing an adsorption efficiency of more than 95% for removal of As(V) from contaminated drinking water. Li et al., (2012b) fabricated polypyrrole/GO (PPy/GO) composite nanosheets by using sacrificial-template polymerization method and applied it to adsorb Cr(VI). The maximum Cr(VI) uptake capacity of PPy/GO composite nanosheets, calculated by fitting the experimental data to the Langmuir equation, was about two times larger than that of conventional PPy nanomaterials. The removal of Pb(II) onto SiO_2/graphene composites synthesized via a two-step reaction, including the preparation of SiO_2/GO and the reduction of GO, was studied by Hao et al., (2012). The composite showed high adsorption efficiency and high selectivity towards Pb(II). The maximum monolayer coverage of Pb(II) on SiO_2/graphene composite was found to be 113.6 mg g^{-1}.

Lee and Yang (2012) demonstrated the preparation of flower-like GO-TiO_2 hybrids for the removal of Zn(II), Cd(II) and Pb(II) from water. The decoration of GO with flower-like TiO_2 nanostructures significantly improved its metal removal efficiency. The adsorption capacities of GO-TiO_2 hybrids, after 6 h and 12 h of hydrothermal treatment at 373 K, were respectively 44.8 ± 3.4 and 88.9 ± 3.3 mg g^{-1} for Zn(II), 65.1 ± 4.4 and 72.8 ± 1.6 mg g^{-1} for Cd(II), and 45.0 ± 3.8 and 65.6 ± 2.7 mg g^{-1} for Pb(II) at pH 5.6. In contrast, pristine GO under identical conditions showed removal capacities of 30.1 ± 2.5, 14.9 ± 1.5, and 35.6 ± 1.3 mg g^{-1} for Zn(II), Cd(II) and Pb(II), respectively.

Luo et al., (2013) determined the adsorption of arsenic by nanocomposites of GO-hydrated zirconium oxide (GO–$ZrO(OH)_2$). Arsenic adsorption was found to be pseudo-second-order with very fast adsorption rate; more than 95% of As(III) and As(V) were removed in the first 10 min, and adsorption equilibrium was established within 15 min. Further, the equilibrium data agreed well with Langmuir's model, and high adsorption capacities of 95.5

and 84.89 mg g^{-1} were obtained for As(III) and As(V), respectively. GO–$ZrO(OH)_2$ could lower arsenic concentration to 0.002 mg L^{-1} (a value much below the maximum contaminant level for drinking water) even at a low dosage of 0.5 g L^{-1}. The presence of other co-existing ions such as NO_3^-, SO_4^{2-}, HCO_3^-, Cl^-, and F^- showed no significant interference on the adsorption of As(III) and As(V) even at concentrations as high as 100 mg L^{-1}. In addition, GO–$ZrO(OH)_2$ exhibited excellent regeneration ability suggesting its use in removal of arsenic from drinking water.

Adsorption of U(VI) from aqueous solution on GO@sepiolite composites was explored by Cheng et al., (2013) in batch experiments. The adsorption was found to be exothermic in nature and followed the Langmuir isotherm. A maximum adsorption capacity of 161.29 mg g^{-1} was recorded at pH 5.0 and a temperature of 298 K. Algothmi et al., (2013) reported the successful removal of Cu(II) from its aqueous solutions using calcium alginate/GO (Ca-Alg_2/GO) hybrid composite gel beads as adsorbent. The Cu(II) uptake equilibrium was best described by the Langmuir isotherm with maximum adsorption of 60.2 mg g^{-1}. The pseudo-second-order rate equation provided a very good fitting to the experimental kinetic data. Peng et al., (2013) prepared GO–FeOOH composites for efficient removal of As(V) from water. The nanocomposite showed excellent adsorption properties with fast adsorption rate and maximum adsorption capacity of 73.42 mg g^{-1}.

Ren et al., (2011) synthesized graphene nanosheet/δ-MnO_2 (GNS/MnO_2) composite by a microwave-assisted method. The adsorption equilibrium, kinetics and thermodynamics of Ni(II) onto GNS/MnO_2 was investigated in batch experiments. The adsorption process was found to follow the Langmuir model and was endothermic in nature. The adsorption kinetic data were best described by the pseudo-second-order rate expression. GNS/MnO_2 exhibited a maximum Ni(II) adsorption capacity of 66.01 mg g^{-1} which was significantly higher than that of MnO_2 as well as pristine graphene nanosheets. Furthermore, 0.1 M HCl was found suitable to effectively regenerate GNS/MnO_2 for further use, with only about 9% loss in initial adsorption capacity. Ren et al., (2012) also studied GNS/MnO_2 for removal of Cu(II) and Pb(II). Similar to their previous investigation (2011), the adsorption process of Cu(II) and Pb(II) were found to follow the Langmuir model with pseudo-second-order kinetics. The maximum adsorption capacity of GNS/MnO_2 was calculated to be around 1637.9 and 793.65 μmol g^{-1} for Cu(II) and Pb(II), respectively. Finally, desorption of both Cu(II) and Pb(II) from the loaded GNS/MnO_2 composite was accomplished by using HCl.

He et al. (2011) investigated the removal of Pb(II) onto porous GO/chitosan (PGOC) materials. The adsorption capacity of PGOC for Pb(II) was found to be 99 mg g^{-1}. Zhang et al., (2011) prepared highly porous biodegradable chitosan–gelatin/GO (C-G/GO) monoliths by a unidirectional freeze-drying method and used it as adsorbent for Cu(II) and Pb(II). The monoliths exhibited good metal uptake capacity and could also be recycled several times without any significant loss in adsorption capacity.

A simple chemical bonding method to synthesize magnetic cyclodextrin–chitosan/GO (CCGO) composite has recently been reported by Li et al., (2013a). The adsorption behaviour of Cr(VI) onto CCGO composite was systematically investigated. The optimum pH for Cr(VI) removal was found to be 3.0. The Langmuir isotherm model exhibited better fit to the adsorption equilibrium data than the Freundlich model implying monolayer coverage of Cr(VI) ions onto the composite surface. The maximum adsorption capacity was found to be 67.66 mg g^{-1} at 323 K. The adsorption followed the pseudo-second-order kinetics. The metal loaded CGCO composite could easily be separated from the treated water because of its magnetic properties, high surface area, and abundant hydroxyl and amino functional groups on its surface. Desorption studies revealed that the spent adsorbent could be regenerated and reused, with less than 5% decrease in adsorption capacity after five adsorption-desorption cycles. In another recent study, Fan et al., (2013a) reported the use of magnetic chitosan/GO (MCGO) materials as an excellent adsorbent for Pb(II). A high Pb(II) adsorption capacity of 76.94 mg g^{-1} and an extremely high desorption capacity of up to 90.3% was obtained, suggesting the potential application of MCGO composite in the removal of Pb(II) from agricultural and industrial wastewaters.

Chandra et al., (2010) synthesized Fe_3O_4-rGO composites with different magnetic concentration to remove arsenic from water. The composites showed near complete (over 99.9%) arsenic removal within 1 ppb and showed high binding capacity for As(III) and As(V). The equilibrium data were found to fit the Langmuir model better than the Freundlich model. Further, the adsorption of As(III) and As(V) by Fe_3O_4-rGO composites showed pseudo-second-order kinetic behaviour.

Yang et al., (2012) decorated both GO and rGO with iron oxide nanoparticles and studied the adsorption of Pb(II) on them separately. They found that the performance of GO-iron oxide composite was significantly better than that of rGO-iron oxide composite, and showed a Langmuir monolayer adsorption capacity of 588.24 mg g^{-1}. Pb(II) adsorption on GO-iron oxide hybrids was strongly dependent on solution pH and independent of ionic strength. Meanwhile, the pH values of the solution before and after adsorption were measured. The pH values after adsorption showed a minor shift to the acidic region implying that surface complexation of Pb(II) with the oxygen functionalities on GO-iron oxide composites was mainly responsible for the high adsorptive uptake of Pb(II).

Magnetic graphene nanocomposites (MGNCs) decorated with core@double-shell nanoparticles (composed of crystalline iron core, iron oxide inner shell and amorphous Si—S—O compound outer shell) were prepared by Zhu et al., (2011) via a facile thermo-decomposition process and used to adsorb Cr(VI). The adsorption process followed pseudo-second-order kinetics with a very fast adsorption rate. Nearly 100% removal was achieved within 5 min using an adsorbent dose of 3 g L^{-1}. Maximum Cr(VI) removal was observed under acidic conditions when the pH was between 1 and 3. Due to the significantly reduced treatment time, MGNCs were considered to be one of the most promising candidates for the efficient removal of heavy metals from wastewater. Recently, Zong et al., (2013) synthesized Fe_3O_4/GO (Fe_3O_4/GO) nanocomposites and employed it as a novel adsorbent for removal of U(VI) from aqueous solution. The maximum adsorption capacity of U(VI) on Fe_3O_4/GO at 293 K and pH = 5.5 ± 0.1, calculated by the Langmuir isotherm, was about 69.49 mg g^{-1}.

The ability of polypyrrole–rGO (PPy/rGO) composites to remove Hg(II) from aqueous solutions was investigated by Chandra and Kim (2011). A high Hg(II) removal capacity of 979.54 mg g^{-1} was obtained. The removal of Hg(II) onto a variety of rGO-metal/metal oxide composites (rGO-MnO_2, rGO-Ag) was studied by Sreeprasad et al., (2011). rGO-MnO_2 and rGO-Ag were found to have nearly similar Hg(II) adsorption capacities of 9.50 and 9.53 mg g^{-1}, respectively. Additionally, a methodology was developed to immobilize rGO-composites on river sand using chitosan as binder. The as-supported composites were found to be efficient adsorbent candidates for field application. Bhunia et al., (2012) successfully synthesized iron–iron oxide matrix dispersed on rGO (rGO–Fe(0)–Fe_3O_4) and used it as adsorbent for the removal of As(III) from aqueous solutions. The adsorption process was found to follow the Langmuir model with pseudo-second-order kinetics. The maximum monolayer coverage of As(III) on rGO–Fe(0)–Fe_3O_4 nanocomposite was 44.4 mg g^{-1}. They also compared the As(III) adsorption behaviour of rGO–Fe(0)–Fe_3O_4 nanocomposite with that of rGO–Fe(0) and rGO–Fe_3O_4 composites and found that rGO–Fe(0)–Fe_3O_4 was the most efficient in adsorbing As(III) from water. Due to its excellent adsorption potential, the removal of some other metal ions by rGO–Fe(0)–Fe_3O_4 was also investigated.

Table 2. Reported results of batch adsorption studies on the removal of dyes from water by graphene and its related materials

Adsorbent	Adsorbate	Conc.	pH	Temp. (K)	Contact time (h)	Adsorption capacity	Reference
Graphene	Cationic Red X-GRL	20-140 mg L^{-1}	—	333	24	238.10 mg g^{-1}	Li et al., 2011a
Graphene	Methyl Blue	5 mg L^{-1}	—	303	96	1.52 g g^{-1}	Wu et al., 2011
Graphene	Methylene Blue	20-120 mg L^{-1}	—	333	—	204.08 mg L^{-1}	Liu et al., 2012a
Graphene sponge	Methylene Blue	2×10^{-4} mol L^{-1}	—	298	4	184 mg g^{-1}	Zhao et al., 2012a
Graphene sponge	Rhodamine B	2×10^{-4} mol L^{-1}	—	298	4	72.5 mg g^{-1}	Zhao et al., 2012a
Graphene sponge	Methyl Orange	2×10^{-4} mol L^{-1}	—	298	24	11.5 mg g^{-1}	Zhao et al., 2012a
GO	Methylene Blue	0.188-1.000 g L^{-1}	6.0	298	1	714 mg mg^{-1}	Yang et al., 2011
GO	Methylene Blue	10-50 mg L^{-1}	10.0	—	—	17.3 mg g^{-1}	Ramesha et al., 2011
GO	Methyl Violet	10-50 mg L^{-1}	6.0	—	—	2.47 mg g^{-1}	Ramesha et al., 2011
GO	Rhodamine B	1-10 mg L^{-1}	6.0	—	—	1.24 mg g^{-1}	Ramesha et al., 2011
GO	Methyl Green	—	9.0	398	1	7.613 mmol g^{-1}	Sharma and Das, 2013
GO	Acridine Orange	0.1 g L^{-1}	—	Room temp.	3	1428 mg g^{-1}	Sun et al., 2012
Sodium hydrosulfite reduced GO	Acridine Orange	0.1 g L^{-1}	—	Room temp.	3	3333 mg g^{-1}	Sun et al., 2012
rGO	Orange G	1-60 mg L^{-1}	—	—	—	5.98 mg g^{-1}	Ramesha et al., 2011
rGO-based hydrogel	Methylene Blue	0.5-10 mg L^{-1}	6.4	298	2	7.85 mg g^{-1}	Tiwari et al., 2013
rGO-based hydrogel	Rhodamine B	0.5-10 mg L^{-1}	6.4	298	2	29.44 mg g^{-1}	Tiwari et al., 2013
Graphene/Fe_3O_4	Fuchsine	20-60 mg L^{-1}	6.6 ± 0.2	298	1	89.4 mg g^{-1}	Wang et al., 2011
Graphene/ Magnetite	Methylene Blue	10-25 mg L^{-1}	—	298	—	43.82 mg g^{-1}	Ai et al., 2011a

Adsorbent	Adsorbate	Conc.	pH	Temp. (K)	Contact time (h)	Adsorption capacity	Reference
Magnetic Fe_3O_4@Graphene	Congo Red	—	—	298 ± 052	—	33.66 mg g^{-1}	Yao et al., 2012a
Magnetic Fe_3O_4@Graphene	Methylene Blue	—	—	298 ± 0.5	—	45.27 mg g^{-1}	Yao et al., 2012a
Graphene/Fe_3O_4	Pararosaniline	20-60 mg L^{-1}	6.6 ± 0.2	298	1	198.23 mg g^{-1}	Wu et al., 2013c
Graphene-SO_3H/Fe_3O_4	Safranine T	20-250 mg L^{-1}	6.0	Room temp.	—	199.3 mg g^{-1}	Wang et al., 2013c
Graphene-SO_3H/Fe_3O_4	Neutral Red	20-250 mg L^{-1}	6.0	Room temp.	—	216.8 mg g^{-1}	Wang et al., 2013c
Graphene-SO_3H/Fe_3O_4	Victoria Blue	20-250 mg L^{-1}	6.0	Room temp.	—	200.6 mg g^{-1}	Wang et al., 2013c
$CoFe_2O_4$-functionalized Graphene sheet	Methyl Orange	10 mg L^{-1}	—	—	—	71.54 mg g^{-1}	Li et al., 2011b
Graphene sheets decorated with $CoFe_2O_4$ nanoparticles	Methyl Green	50-400 mg L^{-1}	—	323	—	312.80 mg g^{-1}	Farghali et al., 2013
Graphene sand composite	Rhodamine 6G	5 mg L-1	—	303 ± 2	8	55 mg g-1	Sen Gupta et al., 2012
Graphene-Sand	Rhodamine 6G	—	—	303 ± 2	6	75.4 mg g-1	Sreeprasad et al., 2013
Graphene-CNT	Methylene Blue	10-30 mg L-1	—	—	3	81.97 mg g-1	Ai and Jiang, 2012
Graphene/c-MWCNT	Rhodamine B	20 mg L-1	—	Room temp.	—	150.2 mg g-1	Sui et al., 2012

Table 2. (Continued)

Adsorbent	Adsorbate	Conc.	pH	Temp. (K)	Contact time (h)	Adsorption capacity	Reference
Graphene/c-MWCNT	Methylene Blue	20 mg L^{-1}	—	Room temp.	—	191.0 mg g^{-1}	Sui et al., 2012
Graphene/c-MWCNT	Fuchsine	20 mg L^{-1}	—	Room temp.	—	180.8 mg g^{-1}	Sui et al., 2012
Graphene-Chitosan composite hydrogel	Methylene Blue	0-80 mg L^{-1}	6.5	294 ± 1	58	390 mg g^{-1}	Chen et al., 2013a
Graphene-Chitosan composite hydrogel	Eosin Y	0-80 mg L^{-1}	7.0	294 ± 1	36	326 mg g^{-1}	Chen et al., 2013a
Fe_3O_4-SiO_2-GO	Methylene Blue	—	—	333	—	111.1 mg g^{-1}	Yao et al., 2012b
Magnetic chitosan/GO	Methyl Blue	60-200 mg L^{-1}	5.3	303	1	95.31 mg g^{-1}	Fan et al., 2012a
Magnetic chitosan/GO	Methylene Blue	50-100 mg L^{-1}	—	303 ± 0.2	—	180.83 mg g^{-1}	Fan et al., 2012b
Polystyrene @Fe_3O_4@GO	Rhodamine B	0-150 mg L^{-1}	—	Room temp.	24	13.8 mg g^{-1}	Wang et al., 2012a
Magnetic β-cyclodextrin-chitosan/GO	Methylene Blue	—	—	298	—	84.32 mg g^{-1}	Fan et al., 2013b
rGO@ZnO	Rhodamine B	4-750 mg L^{-1}	—	293	—	32.6 mg g^{-1}	Wang et al., 2012b
Magnetite/rGO	Rhodamine B	0.5-4 mg L^{-1}	7.0	298	2	13.15 mg g^{-1}	Sun et al., 2011
Magnetite/rGO	Malachite Green	0.5-4 mg L^{-1}	7.0	298	2	22.0 mg g^{-1}	Sun et al., 2011

The maximum adsorption capacity of rGO–Fe(0)–Fe_3O_4 for removal of Cr(VI), Hg(II), Pb(II) and Cd(II) was found to be 31.1, 22.0, 19.7 and 1.91 mg g^{-1}, respectively.

2.1.2. Adsorption of Organic Dyes

Graphene, and its related materials including GO and rGO, as well as their nanocomposites have all been widely examined as a promising class of adsorbents for removal of organic dyes from aqueous environments (Table 2). Li et al., (2011a) reported the use of graphene as an adsorbent for removal of Cationic Red X-GRL from its aqueous solution. The adsorption capacity decreased with increasing adsorbent dose while it increased with increasing temperature. The maximum adsorption capacity increased from 153.85 mg g^{-1} to 204.08 mg g^{-1} with the rise in temperature from 293 K to 333 K. The pseudo-second-order kinetic mechanism was found to be predominant. Adsorption equilibrium data fitted well to the Langmuir isotherm model than the Freundlich model. Estimation of the thermodynamic parameters indicated that adsorption of Cationic Red X-GRL onto graphene was a spontaneous, endothermic and physisorption process.

Adsorption of Methylene Blue onto graphene was studied by Liu et al., (2012a). The maximum dye adsorption capacity obtained from the Langmuir isotherm equation was 153.85 mg g^{-1} at 293 K. Similar to the investigations of Li et al., (2011), the adsorption process was spontaneous and endothermic, and fitted the pseudo-second-order rate equation.

Wu et al., (2011) investigated the use of graphene for removal of Methyl Blue. Graphene showed a remarkably high dye adsorption capacity of 1.52 g g^{-1} which was mainly due to π-π stacking interactions as inferred through fluorescence spectroscopy studies. The researchers also investigated the possibility to regenerate and reuse graphene and observed no significant changes in the efficiency of graphene at least during the first five cycles of the adsorption-desorption process.

In a very interesting study by Zhao et al., (2012a), a new graphene material called "graphene sponge" was developed by hydrothermal treatment of GO sheets in the presence of thiourea. The material had a very porous structure and behaved like a sponge (hence the name). Graphene sponge showed excellent mechanical properties, very good processability and structural stability. The feasibility of using graphene sponge to remove cationic (Methylene Blue, Rhodamine B) and anionic (Methyl Orange) dyes from their aqueous solutions was investigated. Graphene sponge was found to have an affinity for both cationic and anionic dyes. However, the sorption capacity for basic dyes was much higher than for acidic dyes because of the ionic charges on the dyes and surface characteristics of graphene sponge. The maximum adsorption uptake was found to be 184, 72.5 and 11.5 mg g^{-1} for Methylene Blue, Rhodamine B and Methyl Orange, respectively.

GO has also been proved to be an excellent adsorbent for removal of dyes from aqueous systems. For instance, Yang et al., (2011) demonstrated the removal of Methylene Blue from its aqueous solution by using GO as adsorbent. Almost complete removal of Methylene Blue could be achieved at initial dye concentrations of less than 250 mg L^{-1}. The dye removal efficiency increased with increase in pH and ionic strength. Lower temperatures and the presence of dissolved organic matter favored the adsorption process.

Ramesha et al., (2011) evaluated the sorption capacity of GO for the uptake of three different cationic dyes viz. Methylene Blue, Methyl Violet, and Rhodamine B from their aqueous solutions. GO exhibited a significantly high adsorption capacity for all the three dyes. However, the sorption preference increased in the order Methylene Blue > Methyl Violet > Rhodamine B. Such a behavioral pattern was attributed to the fact that Methylene Blue and Methyl Violet were positively charged whereas Rhodamine B had both positive and negative charges associated with its structure and hence the electrostatic interactions between Rhodamine B and GO were considerably weaker. In the same study, the utilization of GO for the removal of an acidic dye, Orange G, was also investigated by Ramesha et al., (2011). However, GO showed a poor binding affinity for Orange G. The two sulfonic groups of Orange G made it negatively charged resulting in electrostatic repulsion between the dye and the adsorbent, and hence no significant removal was observed. GO has also been tested for the removal of Methyl Green dye by Sharma and Das (2013). An adsorption capacity in the range of 4.82—7.61 mmol g^{-1} was recorded at different pH (4.0—9.0).

A study on the use of modified GO for removing Acridine Orange from its aqueous solution was conducted by Sun et al., (2012). In that research, the effectiveness of GO as an adsorbent was attempted to improve through *in situ* reduction with sodium hydrosulfite. Parallel adsorption tests under similar experimental conditions, carried out with pristine GO and *in situ* reduced GO, showed that the later had a much higher adsorption capacity (3333 mg g^{-1}) than the former (1428 mg g^{-1}). It was suggested that the enhancement was probably due to the reduction of carbonyl groups to hydroxyl groups.

The efficacy of rGO in adsorbing dyes has also been explored in recent years. Ramesha et al., (2011) investigated the potential of rGO to remove an acidic dye, Orange G, from its aqueous solutions. An excellent removal efficiency of about 95% was recorded. In another study by Tiwari et al., (2013), a three-dimensional rGO-based hydrogel was synthesized by chemical reduction of GO with sodium ascorbate. The prepared hydrogel had a very large surface area with uniform pore size distribution, and was tested as an adsorbent for the removal of Methylene Blue and Rhodamine B from aqueous solutions in a batch system. The results showed that with an adsorbent dose of 0.6 g L^{-1}, extremely high removal efficiencies of up to ~100% for Methylene Blue and ~97% for Rhodamine B could be achieved within 2 h at room temperature. The high uptake capacity was due to adsorption through strong π-π stacking and anion–cation interactions. Furthermore, desorption studies demonstrated that the rGO-based hydrogel could be easily regenerated by using an inexpensive solvent such as ethylene glycol and reused for at least three adsorption-desorption cycles without any significant loss in adsorption capacity.

Many research studies have recently been undertaken extensively for adsorptive treatment of colored effluents using NGMs. Wang et al., (2011) synthesized a graphene-based magnetite nanocomposite (G/Fe_3O_4) by *in situ* chemical co-precipitation of Fe^{2+} and Fe^{3+} in alkaline solution in the presence of graphene and investigated its potential as an adsorbent for the removal of Fuchsine dye from aqueous solution (Figure 5). The dye uptake process was very fast; about 96% of the dye was adsorbed within 10 min and 99% of the dye was adsorbed within 30 min. The amount of dye adsorbed increased with increase in pH from 3.0 to 5.5. Further increase in pH did not significantly change the adsorption yield. The adsorption equilibrium data was best described by the Langmuir isotherm model, with a maximum monolayer adsorption capacity of 89.4 mg g^{-1}. The adsorption process was found to be a pseudo-second-order reaction. Furthermore, maximum desorption of 94% was achieved at pH 2.0 using ethanol as the eluent. The adsorption capacity of G/Fe_3O_4 for Fuchsine did not show any significant decrease even after five regenerations. Graphene/magnetite composites have also been prepared and successfully employed as adsorbent for the removal of Methylene Blue (Ai et al., 2011a; Yao et al., 2012a), Congo Red (Yao et al., 2012a) and Pararosaniline (Wu et al., 2013c).

Wang et al., (2013c) synthesized a magnetic-sulfonic graphene nanocomposite ($G\text{-}SO_3H/Fe_3O_4$) and explored it as an adsorbent for the batch removal of three cationic dyes: Safranine T, Neutral Red, Victoria Blue, and three anionic dyes: Methyl Orange, Brilliant Yellow, Alizarin Red, from their aqueous solutions. The $G\text{-}SO_3H/Fe_3O_4$ adsorbent showed excellent adsorption capacity towards cationic dyes than anionic dyes. More than 93% of all the cationic dyes were removed during the first 10 min of dye-adsorbent contact. The nanocomposite could be easily regenerated using ethanol (adjusted to pH 2.0 with 0.1 mol L^{-1} HCl) as eluent and reused for at least six adsorption-desorption cycles without any significant loss in adsorption capacity.

Figure 5. Fuchsine dye solution before (right) and after adsorption with G/Fe_3O_4 and separation of the adsorbent with a magnet (left). Reprinted from Wang et al., (2011), Copyright 2011, with permission from Elsevier.

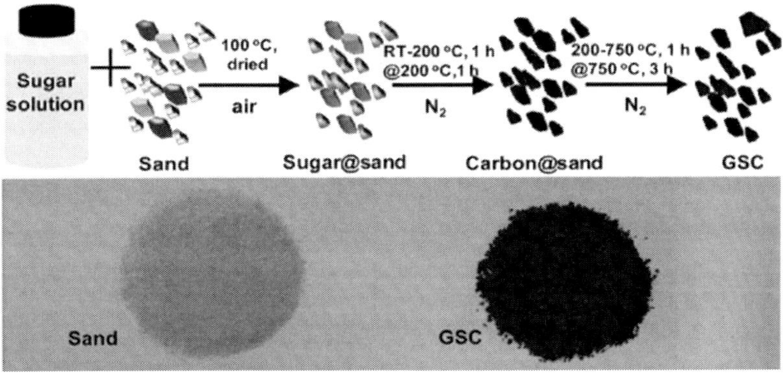

Figure 6. Schematic illustrating the preparation of graphene-sand composite (GSC). Reprinted from Sen Gupta et al. (2012), Copyright 2012, with permission from American Chemical Society.

Magnetic $CoFe_2O_4$-functionalized graphene sheet ($CoFe_2O_4$–FGS) nanocomposites were prepared by Li et al., (2011b) to remove Methyl Orange. The adsorption process followed pseudo-second-order kinetics and the adsorption capacity was found to be as high as 71.54 mg g^{-1}. Recently, Farghali et al., (2013) also synthesized $CoFe_2O_4$–FGS nanocomposites for the removal of Methyl Green from aqueous solution. The adsorption isotherm was well described by the Langmuir model, whereas the adsorption kinetics corresponded to the pseudo-second-order kinetic model. Thermodynamic analyses indicated that adsorption of Methyl Green was spontaneous, endothermic and a physisorption process.

In another study conducted by Sen Gupta et al., (2012), graphene immobilized on sand was used as an adsorbent for the removal of Rhodamine 6G. Graphene was prepared *in situ* from cane sugar and anchored onto the surface of river sand without the need of any additional binder, resulting in a composite, referred to as graphene-sand composite (GSC) (Figure 6). The ability of GSC to remove Rhodamine 6G from its aqueous solution was tested through batch and continuous column experiments. The adsorption process followed pseudo-second-order kinetics and equilibrium was attained in 8 h. The equilibrium adsorption capacity of Rhodamine 6G was 55 mg g^{-1} at 303 ± 2 K. Fixed bed column experiments were performed to study the practical applicability of the adsorbent and breakthrough curves at different bed depths were obtained. The bed depth service time (BDST) model showed good agreement with the dynamic flow experimental data. Desorption studies revealed that GSC could be regenerated using acetone for multiple use. In general, the results suggested that GSC should be considered for dye wastewater treatment with appropriate engineering. The same research group, prepared GSC using asphalt as the carbon source and tested it as an adsorbent for removal of Rhodamine 6G (Sreeprasad et al., 2013). Both batch and continuous column experiments were carried out. Similar to their previous work, the experimental data correlated well with the pseudo-second order model and an adsorption capacity of 75.4 mg g^{-1} was achieved. Fixed-bed column adsorption studies were also conducted in multiple cycles and the results thus obtained confirmed that GSC could be used for water purification applications.

Ai and Jiang (2012) reported the efficient removal of Methylene Blue from its aqueous solution by a self-assembled cylindrical graphene–carbon nanotube hybrid (G-CNT). The hybrid showed good adsorption performance with a maximum adsorption capacity of 81.97 mg g^{-1}. Sui et al., (2012) fabricated graphene–CNT hybrid aerogels by supercritical CO_2 drying of their hydrogel precursors obtained from heating the aqueous mixtures of GO and CNTs with vitamin C without stirring. CNTs used in the study were either pristine MWCNTs or acid-treated MWCNTs (c-MWCNTs). The resulting hybrid aerogels i.e., graphene/MWCNT and graphene/c-MWCNT showed excellent adsorption performance in removal of basic dyes (Rhodamine B,

Methylene Blue, Fuchsine) from their aqueous solutions. The adsorption was found to be pseudo-second-order with the binding capacity of graphene/c-MWCNT (150.2, 191.0 and 180.8 mg g^{-1} for Rhodamine B, Methylene Blue and Fuchsine, respectively) being higher than that of graphene/MWCNT (146.0, 134.9 and 123.9 mg g^{-1} for Rhodamine B, Methylene Blue and Fuchsine, respectively).

The adsorption of Rhodamine B by core–shell structured polystyrene-Fe_3O_4-GO nanocomposites was studied by Wang et al., (2012a). The maximum adsorption capacity was found to be 13.8 mg g^{-1}. The same research group also investigated the removal of Rhodamine B onto rGO/ZnO composite (Wang et al., 2012b). The composite showed an excellent recycling performance for dye adsorption with up to 99% recovery over four cycles. In another study conducted by Yao et al., (2012b), Fe_3O_4/SiO_2-GO nanocomposite was synthesized by a covalent bonding technique to remove Methylene Blue. Isotherm data best fitted the Langmuir model with maximum adsorption capacities of 97.0, 102.6, and 111.1 mg g^{-1} at 298, 318, and 333 K, respectively.

Cheng et al., (2012) developed a three-dimensional chitosan—graphene nanocomposite with large specific surface area and unique mesoporosity and used it as an adsorbent to remove Reactive Black 5 from its aqueous solution. A removal efficiency of 97.5% was obtained with an initial dye concentration of 1 mg L^{-1}. The potential of GO-chitosan (GO-CS) composite hydrogel to remove acidic (Eosin Y) and basic (Methylene Blue) dyes from water was explored by Chen et al., (2013a). The equilibrium adsorption capacities were reported to be 390 and 326 mg g^{-1} for Methylene Blue and Eosin Y, respectively. The investigators also reported that GO–CS hydrogel could be used as a column packing material to fabricate a continuous water purification process.

A novel magnetic chitosan-GO (MCGO) nanocomposite has been developed by covalently binding chitosan on the surface of Fe_3O_4 nanoparticles, followed by covalent functionalization of GO with magnetic chitosan, by Fan et al., (2012a). Simple batch adsorption experiments were conducted to estimate the adsorption properties of MCGO for Methyl Blue. The linearized Langmuir isotherm model best represented the experimental equilibrium data. A maximum monolayer adsorption capacity of 95.31 mg g^{-1} was recorded. The dye uptake kinetics followed a pseudo-second-order mechanism. The values of thermodynamic parameters indicated spontaneous and exothermic nature of the adsorption process. MCGO could be easily regenerated using 0.5 mol L^{-1} NaOH and the adsorption capacity was about

90% of the initial saturation adsorption capacity after four adsorption-desorption cycles. In a separate study, Fan et al., (2012b) found that MCGO also has extraordinary adsorption capacity and fast removal rate for Methylene Blue. The maximum monolayer adsorption capacity was found to be 180.83 mg g^{-1}. The adsorption of Methylene Blue by magnetic β-cyclodextrin–chitosan/GO nanocomposites has also been studied by Fan et al., (2013b). In this case, as much as 84.32 mg g^{-1} of Methylene Blue could be adsorbed as determined by the Langmuir model.

rGO-based nanocomposites have also been developed and investigated for their dye removal potential. Sun et al., (2011) prepared magnetite/rGO (MrGO) nanocomposites for the removal of Rhodamine B and Malachite Green dyes from aqueous solutions. The MrGO nanocomposite exhibited excellent dye removal efficiency, with over 91% of Rhodamine B and over 94% of Malachite Green being removed within 2 h of dye-MrGO contact at room temperature. Desorption studies showed that by using an inexpensive eluent such as ethylene glycol, MrGO could be subjected to multiple rounds of recycle and reuse without any significant change in its adsorption efficiency. In order to further evaluate the practical applicability of the prepared adsorbent material, real water samples, including local industrial waste water and lake water collected from Lake Tai in China, were first contaminated with dyes and then treated using the as-prepared MrGO. It was found that real water samples had little interference with the decolourization efficiency of MrGO. In addition, MrGO also showed excellent removal efficiency for other dye pollutants including Crystal Violet and Methylene Blue from industrial waste water.

1.2.1.3. Adsorption of other organic pollutants

Apart from organic dyes, the adsorption of different other organic pollutants by graphene-based materials has also been the subject of many recent research studies (Table 3). Li et al., (2012c) studied the use of graphene for removal of phenol from aqueous solution. The adsorption equilibrium data showed an excellent fit to both Langmuir and Freundlich isotherm models. The maximum monolayer adsorption capacity was found to be 53.19 mg g^{-1} at pH 6.3, initial phenol concentration = 50 mg L^{-1}, adsorbent dose = 0.5 g L^{-1}, temperature = 333 K and contact time = 48 h. The adsorption kinetics followed the pseudo-second-order kinetic model while a thermodynamic assessment indicated endothermic and spontaneous nature of adsorption of phenol onto graphene.

Laboratory scale batch adsorption experiments were conducted by Xu et al., (2012) to investigate the use of graphene for removal of bisphenol A (BPA) from aqueous solution. The adsorption capacity of graphene for BPA showed no significant variation in a wide pH range of 2.0-7.0, and remained around 87 mg g^{-1}. Beyond pH 7.0, the BPA uptake capacity decreased sharply and reached a minimum of about 30 mg g^{-1} at pH 11. The adsorption kinetic data conformed to the pseudo-second-order kinetic model while the adsorption equilibrium data well fitted to the Langmuir isotherm model. The maximum adsorption capacity of graphene for BPA, calculated from the linearized Langmuir model equation, was 181.82 mg g^{-1} at 302.15 K. Xu et al., (2012) explained that noncovalent interactions such as hydrogen bonding and π-π interactions were essentially responsible for the high adsorption uptake of BPA by graphene. The adsorption process was found to be spontaneous and exothermic in nature. Finally, it was concluded that graphene can be considered as a promising adsorbent for removal of BPA. Wu et al., (2011) also inferred that graphene can be successfully employed for the removal of phenolic compounds, namely acrylonitrile, *p*-toluenesulfonic acid and 1-napthalenesulfonic acid.

The removal of 1-napthol from its aqueous solution by sulfonated graphene nanosheets, prepared from GO, has been examined by Zhao et al., (2011c). A maximum adsorption capacity of 6.4 mmol g^{-1} was recorded at 333.15 K. Strong π-π interactions between graphene sheets and the aromatic organic molecules were largely responsible for the high uptake of 1-naphthol by the sulfonated graphene nanosheets. The adsorption equilibrium data conformed to the Freundlich model while the kinetic data better correlated with the Elovich model than the pseudo-second-order model. Values of ΔG^0 (−10.04 to −14.24 kJ mol^{-1}) and ΔH^0 (20.75 kJ mol^{-1}) suggested the process to be spontaneous and endothermic.

Pei et al., (2013) investigated the feasibility of using graphene as well as GO for the removal of 1,2,4-trichlorobenzene, 2,4,6-trichlorophenol, 2-naphthol and naphthalene from aqueous streams. The results showed that graphene had nearly similar adsorption capacity for all the four organic contaminants at pH 5.0 while the adsorption capacity increased in the order naphthalene < 1,2,4-tricholorobenzene < 2,4,6-trichlorophenol < 2-napthol for GO.

Table 3. Reported results of batch adsorption studies on the removal of organic pollutants from water by graphene and its related materials

Adsorbent	Adsorbate	Conc.	pH	Temp. (K)	Contact time (h)	Adsorption capacity	Reference
Graphene	Phenol	10-60 mg L^{-1}	6.3	333	48	53.19 mg g^{-1}	Li et al., 2012c
Graphene	Acrylonitrile	5 mg L^{-1}	—	303	96	0.72 g g^{-1}	Wu et al., 2011
Graphene	p-Toluenesulfonic acid	5 mg L^{-1}	—	303	96	1.43 g g^{-1}	Wu et al., 2011
Graphene	1-Napthalenesulfonic acid	5 mg L^{-1}	—	303	96	1.52 g g^{-1}	Wu et al., 2011
Graphene	Bisphenol A	—	6.0	302.15	6	181.82 mg g^{-1}	Xu et al., 2012
Sulfonated graphene	1-Napthol	0.08 g L^{-1}	7.0	333.15	—	6.4 mmol g^{-1}	Zhao et al., 2011c
GO	Tetracycline	8.33-333.33 mg L^{-1}	3.6	298	Overnight	313.48 mg g^{-1}	Gao et al., 2012a
GO	Oxytetracycline	8.33-333.33 mg L^{-1}	3.6	298	Overnight	212.31 mg g^{-1}	Gao et al., 2012a
GO	Doxycycline	8.33-333.33 mg L^{-1}	3.6	298	Overnight	398.41 mg g^{-1}	Gao et al., 2012a
GO	Tetrabromobisphenol A	0.3-1.0 g L^{-1}	6.0	288	10	132.25 mg g^{-1}	Zhang et al., 2013
Fe_3O_4/Graphene	Aniline	—	—	318	1	375.94 mg g^{-1}	Chang et al., 2012
Fe_3O_4/Graphene	*p*-Chloroaniline	—	—	318	1	43.86 mg g^{-1}	Chang et al., 2012
GO-Iron oxide	1-Napthol	5-75 mg L^{-1}	6.5 ± 0.1	303	48	228.41 mg g^{-1}	Yang et al., 2012

Table 3. (Continued)

Adsorbent	Adsorbate	Conc.	pH	Temp. (K)	Contact time (h)	Adsorption capacity	Reference
GO-iron oxide	1-Napthylamine	5-75 mg L^{-1}	6.5 ± 0.1	303	48	285.71 mg g^{-1}	Yang et al., 2012
rGO-iron oxide	1-Napthol	5-75 mg L^{-1}	6.5 ± 0.1	343	48	588.23 mg g^{-1}	Yang et al., 2012
rGO-iron oxide	1-Napthylamine	5-75 mg L^{-1}	6.5 ± 0.1	343	48	625.00 mg g^{-1}	Yang et al., 2012
rGO/FeO• Fe_2O_3	1-Napthylamine	0.1 g L^{-1}	7.0 ± 0.1	283.15	48	2.85 mmol g^{-1}	Yang et al., 2013
rGO/FeO• Fe_2O_3	1-Napthol	0.1 g L^{-1}	7.0 ± 0.1	283.15	48	2.70 mmol g^{-1}	Yang et al., 2013
rGO/FeO• Fe_2O_3	Napthalene	0.1 g L^{-1}	7.0 ± 0.1	323.15	48	5.72 mmol g^{-1}	Yang et al., 2013
Fe_3O_4@ GO	Polychlorinated Biphenyl 28	1.47 mg L^{-1}	—	—	—	0.718 mg g^{-1}	Zeng et al., 2013
rGO/Magnetite	Ciprofloxacin	1-10 mg L^{-1}	6.2	298	—	18.22 mg g^{-1}	Tang et al., 2013a
rGO/Magnetite	Norfloxacin	1-10 mg L^{-1}	6.2	298	—	22.20 mg g^{-1}	Tang et al., 2013a

The effectiveness of GO as an adsorbent for the removal of tetracycline antibiotics, namely tetracycline, oxytetracycline and doxycycline, from aqueous solutions was estimated by Gao et al., (2012a) using a batch procedure. The maximum amount of antibiotics adsorbed at equilibrium was 313.48, 212.31, and 398.41 mg g^{-1} for tetracycline, oxytetracycline and doxycycline, respectively. The sorption capacities were found to decrease with increase in pH as well as ionic strength. The adsorption equilibrium followed both Langmuir and Temkin isotherm models while the adsorption dynamics showed pseudo-second-order kinetic behavior.

Zhang et al., (2013) reported the removal of tetrabromobisphenol A (TBBPA) from its aqueous solution by GO. The effects of contact time, pH, temperature, and the presence of coexisting anions as well as humic acid, were studied through batch experiments. The adsorption capacity decreased with an increase in temperature as well as solution pH. The presence of coexisting anions such as NO_3^-, SO_4^{2-}, HPO_4^{2-} and HCO_3^{2-} reduced TBBPA sorption in the order $NO_3^- < SO_4^{2-} < HPO_4^{2-} < HCO_3^{2-}$. The presence of humic acid also significantly reduced the adsorption of TBBPA onto GO. The adsorption reaction followed the pseudo-second-order kinetics, while the adsorption isotherm was well described by the Langmuir model with a maximum adsorption capacity of 132.25 mg g^{-1} at 288 K. Thermodynamic studies indicated that the adsorption of TBBPA onto GO was feasible and spontaneous at all temperatures and was exothermic in nature. Both π-π stacking interactions and hydrogen bonding were proposed to be responsible for the adsorption of TBBPA by GO.

Recently, Pavagadhi et al., (2013) have shown that GO can be utilized as an adsorbent of algal toxins, microcystin-LR (MC-LR) and microcystin-RR (MC-RR). The adsorption performance of GO was compared to that of commercially available activated carbon. GO showed a very high adsorption capacity of 1700 μg g^{-1} for MC-LR and 1878 μg g^{-1} for MC-RR while the maximum adsorption capacity obtained with commercial activated carbon was 1481.7 and 1034.1 μg g^{-1} for MC-LR and MC-RR, respectively. The sorption kinetic experiments revealed that more than 90% removal of both MC-LR/RR was achieved within 5 min for all the doses studied (500, 700 and 900 μg L^{-1}). GO could be reused as an adsorbent following ten cycles of adsorption/desorption with no significant loss in its adsorption capacity.

A few researchers have also focused on the development of NGMs to adsorb organic pollutants from wastewater. Chang et al., (2012) investigated the adsorption performance of Fe_3O_4/graphene for aniline and *p*-chloroaniline. Optimum removal was observed in the pH range 5.0-9.0 (Figure 7), while a shaking time of 60 min was sufficient to reach equilibrium. The pseudo-second-order kinetic model best described the adsorption kinetics of aniline and *p*-chloroaniline on Fe_3O_4/graphene nanocomposite. The Freundlich isotherm model showed a good fit to the equilibrium adsorption data implying that the adsorption of aniline and *p*-chloroaniline on Fe_3O_4/graphene nanocomposite was multilayer and applicable to heterogeneous surfaces.

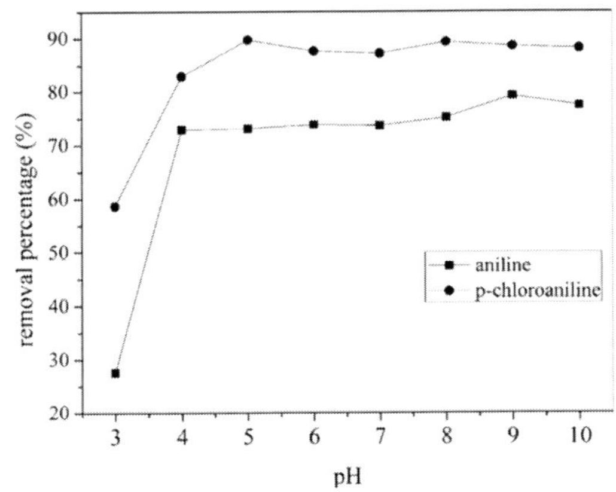

Figure 7. Effect of initial solution pH on percentage removal of aniline and p-chloroaniline by Fe_3O_4/graphene composite adsorbent. Reprinted from Chang et al., (2012), Copyright 2012, with permission from Elsevier.

Yang et al., (2012) decorated both GO and rGO with iron oxide nanoparticles and explored them as adsorbents for removal of organic pollutants (1-naphthylamine and 1-naphthol). rGO/iron oxide material was found to be a better adsorbent for 1-naphthol and 1-naphthylamine than GO/iron oxide hybrid material. The adsorption of organic pollutants on rGO/iron oxide nanocomposites was found to be an endothermic and spontaneous process. Inspired by the enormous potential of rGO-based composite materials for treatment of organic pollutants, the same research group also prepared rGO/iron oxide (rGO/$FeO{\bullet}Fe_2O_3$) composites and used it as a super adsorbent for removal of 1-naphthylamine, 1-naphthol and naphthalene from their aqueous solution (Yang et al., 2013). The adsorption rate of napthalene and its derivatives by the as-synthesized rGO/$FeO{\bullet}Fe_2O_3$ composites followed the order, 1-naphthylamine>1-naphthol>napthalene. The adsorption behavior followed the Freundlich adsorption isotherm model. Yang et al., (2013) also investigated the adsorption mechanism in detail and postulated that π-π EDA interaction (between benzene ring of the aromatic compounds and rGO/FeO•Fe_2O_3 surface) was the primary adsorption mechanism involved in the removal of organic compounds.

Zeng et al., (2013) reported the development and use of Fe_3O_4 nanoparticle grafted GO (Fe_3O_4@GO) nanocomposite for the removal of polychlorinated biphenyl 28. The adsorption kinetics was found to fit the pseudo-second-order rate equation better than Lagergren's first-order model. The results suggested that Fe_3O_4@GO could be considered as a suitable material for the abatement of polychlorinated biphenyl pollution.

Recently, the adsorption of fluoroquinolone antibiotics, namely ciprofloxacin (CIP) and norfloxacin (NOR), by rGO/magnetite (rGO-M) composites have been investigated by Tang et al., (2013a). The results of batch equilibrium tests indicated that CIP and NOR adsorption on rGO-M was strongly dependent on solution pH, and involved π-π interactions as well as electrostatic repulsions. The adsorption equilibrium data were in good agreement with the Langmuir and Temkin isotherm models. The maximum monolayer uptake of CIP and NOR onto rGO-M was found to be 18.22 and 22.20 mg g^{-1}, respectively. Adsorption of CIP and NOR on rGO-M could be well described by the pseudo-second-order kinetic model. The thermodynamic parameters indicated that the adsorption of fluoroquinolone antibiotics by rGO-M was spontaneous and exothermic.

2.2. Water Remediation by Heterogeneous Photocatalysis

In recent years, semiconductor-mediated heterogeneous photocatalysis has also emerged as one of the most powerful methods of water decontamination because of its potential to transform recalcitrant organic contaminants into mineral salts and relatively innocuous end products such as CO_2 and H_2O (Augugliaro et al., 2012; Sud and Kaur, 2012; Wang and Xu, 2012). This green technology offers a number of advantages such as ambient operating conditions, low operating costs, and complete mineralization of organic pollutant without any secondary pollution, all of which have promoted its widespread application in wastewater treatment (Ahmed et al., 2011). Unfortunately, the insufficient quantum efficiency, narrow excitation wavelength, high recombination rate of the photoproduced electron-hole pairs, poor adsorption capacity, and deactivation of the semiconductor photocatalyst limit the practical application of this technique (Han et al., 2012; Zhang et al., 2012a).

Graphene materials, due to their high adsorption capacity, large surface area, extended light absorption range, enhanced charge separation and charge transport properties, are an ideal electronic sink or electron-transfer bridge and a 2D support matrix for photocatalyst carrier or promoter (Leary and Westwood, 2011; Han et al., 2012; Zhang et al., 2012a). There has, therefore, been an increasing research interest to couple graphene with photoactive semiconductors to develop graphene/semiconductor nanocomposites (GSNs) with high photocatalytic performance towards pollution control and abatement. Till date, a wide range of semiconductor materials have been supported on graphene-based templates. These materials mainly include metal oxides (e.g., TiO_2 [Zhou et al., 2011a; Huimin et al., 2012; Zhao et al., 2012b; Ismail et al., 2013], ZnO [Xu et al., 2011; Herring et al., 2012; Ahmad et al., 2013; Wei et al., 2013], Cu_2O [Gao et al., 2012b], Fe_2O_3 [Chen et al., 2013b], Mn_2O_3 [Chandra et al., 2012], WO_3 [Zhou et al., 2012]), metal sulfides (e.g., CdS [Pan and Liu, 2012], In_2S_3 [An et al., 2013], Sb_2S_3 [Tao et al., 2013]), metallates (e.g., $BiVO_4$ [Fu et al., 2011], $CoFe_2O_4$ [Fu et al., 2012]; Bi_2WO_6 [Xu et al., 2013], Bi_2MoO_6 [Zhou et al., 2011b]), and other nanoparticles (e.g., CdSe [Ghosh et al., 2013], $Zn(CH_3COO)_2$ [Pawar et al., 2013]). Various techniques have been developed for preparing such composite photocatalysts of which solution mixing, hydrothermal/solvothermal synthesis, sol-gel method, microwave-assisted deposition and photo-assisted reduction are predominantly used at present (Huang et al., 2012; Wang et al., 2013a). The as-synthesized GSNs have been found to be effective in the photodegradation of organic and inorganic contaminants as well as harmful biological pollutants.

2.2.1. Photocatalytic degradation of inorganic pollutants

Photocatalytic degradation of metal ions by GSNs has been extensively studied by Liu and co-workers (Liu et al., 2011a; Liu et al., 2011b; Liu et al., 2011c). They have shown that TiO_2—rGO composites, synthesized *via* the microwave-assisted reduction of GO in a TiO_2 suspension, exhibited enhanced photocatalytic reduction of Cr(VI) as compared with pure TiO_2 and commercial P25 (Liu et al., 2011a). A maximum removal rate of 91% was achieved under UV light irradiation. In another study, Liu et al., found that ZnO—rGO composites prepared by microwave-assisted reduction of $ZnSO_4$ in an aqueous GO suspension could effectively reduce Cr(VI) when irradiated with UV light as compared with pure ZnO (Liu et al., 2011b). Similarly, CdS—rGO composites have also been found to exhibit enhanced photocatalytic performance than pure CdS for the reduction of Cr(VI) by Liu's group (Liu et al., 2011c). Liu et al., (2011a; 2011b; 2011c) concluded that introduction of graphene material as a substrate increases the light absorption intensity and range and reduces the electron-hole pair recombination which in turn improves the photocatalytic performance.

Zhang et al. (2012b) established that when TiO_2 is linked covalently to GO through an esterification reaction between the —OH of TiO_2 and the —COOH groups on the GO sheets, its photodegradation efficiency for Cr(VI) is significantly increased. Likewise, Ma et al., (2012) reported that rGO—$Bi_{2-}WO_6$ composites can be used in the reduction of Cr(VI) from its aqueous solutions. In another study, An et al., (2013) investigated the photocatalytic efficiency of In_2S_3—rGO composites prepared using a one pot procedure. The In_2S_3—rGO nanocomposite showed excellent photocatalytic activity towards the degradation of Cr(VI).

2.2.2. Photocatalytic Degradation of Organic Dyes

GSNs as a new type of photocatalyst have been widely investigated for the degradation and mineralization of common textile dyes (Table 4). For instance, Zhou et al., (2011a) prepared graphene/TiO_2 (G/TiO_2) composites through a one-pot solvothermal reaction by using GO and tetrabutyl titanate ($C_{16}H_{36}O_4Ti$) as starting materials. The as-synthesized nanocomposite exhibited significantly enhanced photodegradation of Methylene Blue over pure TiO_2 under irradiation with simulated sunlight. Lee et al., (2012) decorated graphene sheets with TiO_2 nanorods by a simple non-hydrolytic sol–gel approach to develop novel photocatalyst systems with improved activity and performance. Compared to pristine TiO_2 nanorods as well as commercial TiO_2 (Degussa P25), the G/TiO_2 nanorod composite showed unprecedented photodecomposition efficiency (~100%) for Methylene Blue under both UV and visible light. In another study, Nguyen-Phan et al., (2011) demonstrated the excellent photocatalytic performance of GO/TiO_2 composite materials under both UV and visible radiation for degradation of Methylene Blue. The authors explained that irrespective of the light sources, the synergistic effect of GO's large surface area and oxygen-containing surface functional moieties improved the adsorption of dye molecules and the separation efficiency of e^- and h^+, thereby enhancing the photocatalytic performance. rGO/TiO_2 composites have also been found to be at least 6 times more effective for photodegradation of Methylene Blue than by Degussa P25, whether it was under UV or visible light illumination (Ismail et al., 2013).

Table 4. Photocatalytic degradation of synthetic dyes by GSNs in aqueous solutions

Photoactive nanocomposite	Dye	Initial dye conc.	Light source	Irradiation time (min)	Degradation (%)	Reference
Graphene/TiO_2	Methylene Blue	1.0×10^{-5} M	Simulated sunlight	480	>75	Zhou et al., 2011a
Graphene–TiO_2 nanorod	Methylene Blue	3.0×10^{-5} M	Visible	180	~100	Lee et al., 2012
Graphene–TiO_2 nanorod	Methylene Blue	3.0×10^{-5} M	UV	20	~100	Lee et al., 2012
Graphene-Bi_2MoO_6	Methylene Blue	1.0×10^{-5} M	Visible	120	80	Zhou et al., 2011b
TiO_2/GO	Methylene Blue	2.5×10^{-5} M	UV	60	100	Nguyen-Phan et al., 2011
TiO_2/GO	Methylene Blue	2.5×10^{-5} M	Visible	60	95	Nguyen-Phan et al., 2011
TiO_2–GO	Methylene Blue	10 mg L^{-1}	Visible	180	>80	Ismail et al., 2013
TiO_2–GO	Methylene Blue	10 mg L^{-1}	UV	12	100	Ismail et al., 2013
rGO–Bi_2WO_6	Methylene Blue	15 mg L^{-1}	Visible	90	93.5	Xu et al., 2013a
$BiVO_4$-Graphene	Methyl Orange	20 mg L^{-1}	Visible	300	99	Fu et al., 2011
BiOBr/Graphene	Methyl Orange	7.5 mg L^{-1}	Visible	140	>80	Song et al., 2012
BiOI–Graphene	Methyl Orange	10 mg L^{-1}	Visible	240	88	Liu et al., 2013
GO/BiOI	Methyl Orange	20 mg L^{-1}	Visible	180	>80	Liu et al., 2012b
rGO/BiOI	Methyl Orange	20 mg L^{-1}	Visible	180	99	Liu et al., 2012b
CdSe–Graphene–TiO_2	Methyl Orange	1.0×10^{-4} M	Visible	180	71	Ghosh et al., 2013
Ag_3VO_4/TiO_2/Graphene	Methyl Orange	10 mg L^{-1}	Visible	180	81	Wang et al., 2013c
BiOBr–Graphene	Rhodamine B	10 mg L^{-1}	Visible	24	100	Tu et al., 2012
rGO/Bi_2WO_6	Rhodamine B	1	Visible	40	98	Ma et al., 2012
rGO/Bi_2WO_6	Rhodamine B	1	UV	75	98	Ma et al., 2012
$Bi_5Nb_3O_{15}$/Graphene	Rhodamine B	1.0×10^{-5} M	Visible	80	>80	Min et al., 2012b
rGO–$CoFe_2O_4$	Rhodamine B	10 mg L^{-1}	Visible	180	>85%	Bai et al., 2012
rGO/$KNbO_3$	Rhodamine B	1	UV	30	68.3	Li et al., 2013b
TiO_2–dx–rGO	Rhodamine B	5 mg L^{-1}	UV	<60	>90	Shi et al., 2012
$BiVO_4$-Graphene	Active Black BL-G	20 mg L^{-1}	Visible	300	99	Fu et al., 2011

Table 4. (Continued)

Photoactive nanocomposite	Dye	Initial dye conc.	Light source	Irradiation time (min)	Degradation (%)	Reference
$CoFe_2O_4$–Graphene	Active Black BL-G	20 mg L^{-1}	Visible	240	66	Fu et al., 2012
$CoFe_2O_4$–Graphene	Active Red RGB	20 mg L^{-1}	Visible	240	61	Fu et al., 2012
Graphene–Bi_2MoO_6	Reactive Brilliant Red X-3B	25 mg L^{-1}	Visible	90	90.4	Wang et al., 2012c
P25–Graphene	Reactive Black 5	10 mg L^{-1}	UV	100	>90	Li et al., 2013c
Ag–ZnO/rGO	Reactive Black 5	10 mg L^{-1}	UV	200	>80	Pant et al., 2013a
CeO_2–TiO_2–Graphene	Reactive Red 195	20 mg L^{-1}	UV	100	~90	Ghasemi et al., 2012
Pt–TiO_2–Graphene	Reactive Red 195	20 mg L^{-1}	UV	120	>75	Ghasemi et al., 2013
Pd–TiO_2–Graphene	Reactive Red 195	20 mg L^{-1}	UV	120	>60	Ghasemi et al., 2013
Graphene/Polyaniline	Rose Bengal	—	Visible	180	56	Ameen et al., 2012
Graphene/BiOBr	Sulforhodamine 640	10 mg L^{-1}	Visible	60	34.8	Zhang et al., 2012c
Graphene–Mn_2O_3	Eosin	2.5×10^{-5} M	UV	140	80	Chandra et al., 2012
ZnO–GO	Crystal Violet	—	UV	80	~95	Ameen et al., 2013
ZnO–rGO	Malachite Green	6.0×10^{-5} M	UV	90	78	Herring et al., 2012
ZnO–Bi_2O_3/GO	Acid Blue	10 mg L^{-1}	Visible	120	98.31	He et al., 2012
ZnO–Bi_2O_3/GO	Acid Yellow	10 mg L^{-1}	Visible	120	88.69	He et al., 2012
ZnO–Bi_2O_3/GO	Acid Red	10 mg L^{-1}	Visible	120	81.93	He et al., 2012
ZnO–Bi_2O_3/GO	Reactive Blue	10 mg L^{-1}	Visible	120	44.23	He et al., 2012
ZnO–Bi_2O_3/GO	Reactive Yellow	10 mg L^{-1}	Visible	120	80.28	He et al., 2012
ZnO–Bi_2O_3/GO	Reactive Red	10 mg L^{-1}	Visible	120	85.35	He et al., 2012

Xu et al., (2011) synthesized ZnO/graphene (ZnO/G) nanocomposites for photocatalytic degradation of Methylene Blue under UV irradiation. The photoactive nanocomposite exhibited an outstanding performance in decomposing Methylene Blue and its activity was improved by a factor of about 4 and 5 relative to the activity of mechanical mixture of ZnO and graphene and pure ZnO, respectively (Figure 8). These results suggested that intimate contact between semiconductor and graphene material is crucial for the formation of electronic interaction and inter-electron transfer at the interface.

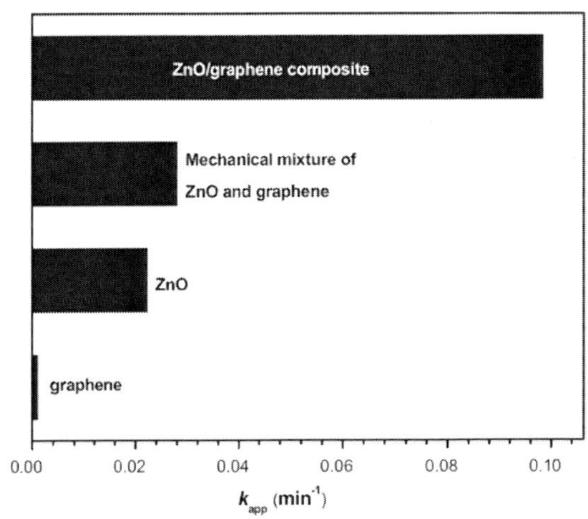

Figure 8. The apparent rate constant of Methylene Blue photodegradation on graphene, ZnO, mechanical mixture of ZnO and graphene, and the ZnO/graphene composite. Reprinted from Xu et al. (2011), Copyright 2011, with permission from Elsevier.

Zhou et al., (2011b) fabricated graphene—γBi_2MoO_6 (G—γBi_2MoO_6) composites by a facile one-step hydrothermal method and examined its photocatalytic performance for degradation of Methylene Blue under visible light. Photodecomposition of Methylene Blue by G—γBi_2MoO_6 followed pseudo-first-order kinetics with a reaction rate constant almost 4 times faster than that of pure Bi_2MoO_6. Xu et al., (2013) compared the photocatalytic activities of pure Bi_2WO_6 and rGO—Bi_2WO_6 composite samples for degradation of Methylene Blue under visible light irradiation. The results suggested that introduction of rGO distinctly enhanced the photoactivity of Bi_2WO_6.

Fu et al., (2011) developed $BiVO_4$–graphene ($BiVO_4$–G) photocatalyst by a facile one-step hydrothermal method without the use of any surfactants. The photocatalytic measurements demonstrated that $BiVO_4$–G nanocomposites were more photoactive than pure $BiVO_4$ photocatalysts for degradation of Methyl Orange. Almost 99% of the dye was degraded by the $BiVO_4$–G nanocomposite after about 300 min exposure to visible light. Song et al., (2012) studied the photocatalytic degradation of Methyl Orange under visible light irradiation in the presence of BiOBr/graphene nanocomposites. The experimental results showed that the photodecolorization efficiency of the composite photocatalyst was significantly higher than that of commercial P25 catalyst. Liu et al., (2013) successfully synthesized a BiOI–graphene composite photocatalyst and used it to degrade Methyl Orange under visible light irradiation. In addition, GO/BiOI and rGO/BiOI nanocomposites have also been reported for the photodegradation of Methyl Orange in aqueous phase (Liu et al., 2012b). However, the photocatalytic efficiency of GO/BiOI was found to be considerably lower than that of rGO/BiOI.

Ghosh et al., (2013) fabricated CdSe–graphene–TiO_2 (CdSe–G–TiO_2) ternary hybrids by calcination of CdSe–graphene composites with titanium (IV) n-butoxide as the source of TiO_2 at 600 °C. After 180 min of exposure to visible light, about 71% of Methyl Orange was mineralized by the as-synthesized CdSe–G–TiO_2 nanocomposites. In addition, there was no significant decrease in the photocatalytic activity of CdSe–G–TiO_2 even after four recycles. The photodegradation followed first-order kinetics with a reaction rate constant of 0.0041 min^{-1}. Wang et al., (2013d) studied the effectiveness of Ag_3VO_4/TiO_2/graphene (Ag_3VO_4/TiO_2/G) nanocomposite for the degradation of Methyl Orange under visible light illumination. The Ag_3VO_4/TiO_2/G composite exhibited a significantly higher photocatalytic activity than Ag_3VO_4/TiO_2 or TiO_2/G composites as well as Ag_3VO_4, nitrogen doped TiO_2 and commercial Degussa P25. The enhanced photoactivity of Ag_3VO_4/TiO_2/G was attributed to the combined effect of formation of hetero-junction between Ag_3VO_4 and TiO_2, electron storage and shuttling ability of graphene and also its wide spectral response.

Tu et al., (2012) reported the complete decolorization of Rhodamine B solutions by visible light induced BiOBr–graphene composites, prepared through a facile *in situ* solvothermal method. Ma and co-workers (Ma et al., 2012) fabricated a Bi_2WO_6/graphene composite that showed excellent photocatalytic activity for the degradation of Rhodamine B under both UV and visible light irradiation. Min et al., (2012) synthesized nanosized bismuth niobate ($Bi_5Nb_3O_{15}$) by a facile hydrothermal method, which were then self-assembled onto graphene sheets to prepare $Bi_5Nb_3O_{15}$/graphene ($Bi_5Nb_3O_{15}$/G) nanocomposites. The photocatalytic decolorization ratio of the nanosized $Bi_5Nb_3O_{15}$ for Rhodamine B was remarkably improved (2.8 times) after coupling with graphene. The novel $Bi_5Nb_3O_{15}$/G composite was also found to have higher photodecolorization efficiency for Rhodamine B compared to Degussa P25. Moreover, the composite also did not show any change in decolorization efficiency even after recycle and reuse for five consecutive cycles.

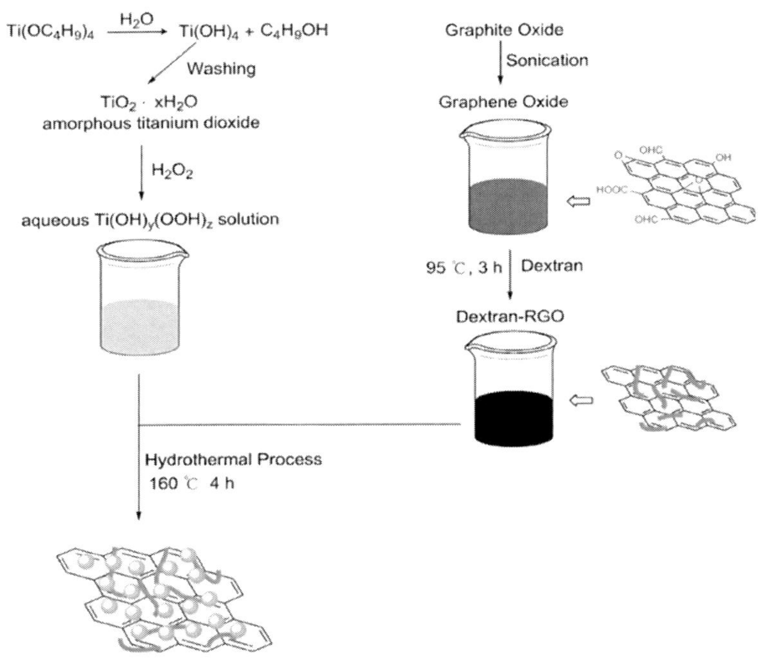

Figure 9. Protocol for the fabrication of TiO_2–dx–rGO nanocomposites. Reprinted from Shi et al., (2012), Copyright 2012, with permission from Elsevier.

Bai et al., (2012) fabricated a series of rGO–MFe_2O_4 (M = Mn, Zn, Co and Ni) hybrids by a one-pot solvothermal method, which were then examined for the visible light decomposition of Rhodamine B. The photocatalytic activity of the different rGO–MFe_2O_4 hybrids decreased in the following order: rGO–$CoFe_2O_4$ > rGO–$ZnFe_2O_4$ > rGO–$MnFe_2O_4$ > rGO–$NiFe_2O_4$. Recently, Li et al., (2013b) prepared rGO/potassium niobate (rGO/$KNbO_3$) composite nanoscrolls by incorporating reduced rGO sheets into $KNbO_3$ sheets at room temperature. The rGO/$KNbO_3$ composite nanoscroll exhibited a moderate Rhodamine B degradation efficiency of 68.3% after UV irradiation for about 30 min.

Another important investigation that deserves special mention is the preparation of TiO_2–dextran–rGO composites by an environmentally friendly and efficient strategy for photodegradation of Rhodamine B (Shi et al., 2012). In this study, GO was first reduced to rGO by using dextran as reducing agent and a surface modifier. TiO_2 nanoparticles were then successfully prepared from a water soluble precursor, peroxotitanium acid and self-assembled on dx—rGO nanosheets through hydrogen bonding and van der Waals interactions (Figure 9). Compared with TiO_2, the TiO_2–dx–rGO composite exhibited remarkably enhanced photocatalytic performance for degradation of Rhodamine B under UV irradiation. The integration of rGO into the matrix of TiO_2 not only increased the light absorption intensity but also improved the adsorptivity of the dye pollutant. Rhodamine B molecules could transfer from the solution to the catalyst surface and be adsorbed with offset face-to-face orientation via π-π conjugation between dye and aromatic regions of rGO. Also, the large surface area of rGO offered adequate active sites to participate in photocatalysis, which ultimately enhanced the photocatalytic activity of TiO_2.

Wang et al., (2012c) demonstrated the synthesis of graphene–Bi_2MoO_6 (G–Bi_2MoO_6) nanocomposite by a one-pot solvothermal method and tested its photocatalytic performance for the degradation of Reactive Brilliant Red X-3B dye under visible light irradiation. Nearly 90.4% degradation of X-3B was achieved after 90 min irradiation, whereas only 63.8% degradation was obtained using pure Bi_2MoO_6 for the same time. The variation in degradation efficiency was attributed to increased migration efficiency of photo-induced electrons as well as increased adsorption affinity owing to the presence of graphene. The degradation process of X-3B followed first-order reaction kinetics with a rate constant of 0.022 min^{-1}. Zhang et al., (2012c) prepared graphene/BiOBr (G/BiOBr) composites by a typical hydrothermal reaction between GO and BiOBr. The presence of graphene on the surface of BiOBr significantly improved the photocatalytic activity of the bulk material for the degradation of Sulforhodamine 640 dye under visible light irradiation. The photocatalytic efficiency of BiOBr was enhanced because of the low isoelectric characteristics of graphene and better interfacial electron transfer between BiOBr and graphene. The pseudo-first-order kinetic model was applied to study the photocatalytic kinetics of Sulforhodamine 640. The high correlation coefficient values (R^2>0.99) obtained by linear regression analysis, confirmed that the photodecomposition process was kinetically controlled by a pseudo-first-order rate law.

Fu et al., (2012) found that $CoFe_2O_4$–graphene ($CoFe_2O_4$–G) nanocomposites fabricated by a facile one-step hydrothermal method were highly efficient and stable than pure $CoFe_2O_4$ for visible light-induced decomposition of Active Black BL-G and Active Red RGB dyes. The increased photoactivity of $CoFe_2O_4$–G was related to the reduction of GO, since the excellent conductivity of the reduced GO sheets was favourable for the efficient separation of photogenerated carriers in the $CoFe_2O_4$ and graphene coupling system. Chandra et al., (2012) investigated the photocatalytic potential of graphene–Mn_2O_3 composites to degrade Eosin. A removal efficiency of 80% was observed within 140 min of UV irradiation. He et al., (2012) demonstrated that ZnO–Bi_2O_3/GO is a very efficient visible light sensitive photocatalyst capable of degrading a variety of textile dyes (such as Acid Blue, Acid Yellow, Acid Red, Reactive Blue, Reactive Yellow, and Reactive Red). Recently, Li et al., (2013c) constructed P25-graphene (P25-G) nanocomposites with high specific surface area and narrow band gap energy that were much more efficient for the UV light degradation of Reactive Black 5 than Degussa P25. Upon 100 min UV irradiation, >90% photodegradation of Reactive Black 5 was observed with P25–G.

Ghasemi and co-workers (Ghasemi et al., 2012) evaluated the photodecomposition of Reactive Red 195 under UV light illumination by TiO_2 and CeO_2–TiO_2 (CT) nanoparticles, and CeO_2–TiO_2–carbon nanotubes (CT–CNTs), CeO_2–TiO_2–activated carbon (CT–AC) and CeO_2–TiO_2–graphene (CT–G) nanocomposites. The order of photocatalytic efficiency was TiO_2 < CT < CT-AC < CT–CNTs < CT–G. The CT–G nanocomposite demonstrated the highest photocatalytic activity due to its unique structure and unprecedented electronic properties. Unexpectedly, the photocatalytic activity of CT–G was found to be inversely proportional to the graphene content in the nanocomposite. In another recent study, Ghasemi et al., (2013) demonstrated that M–TiO_2–graphene (M–TiO_2–G; M = Pt, Pd) composites can also efficiently degrade Reactive Red 195 under both UV and visible light irradiation. However, the photocatalytic activity under UV light was much higher than visible light. The visible-light irradiation with λ >400 nm did not had sufficient energy to excite the valence band electrons in high extent, resulting in lesser number of excited electrons and hence lower degradation compared to UV light. Meanwhile, the photoactivity of Pt–TiO_2–G was found to be higher than that of Pd–TiO_2–G due to high photonic efficiency of the nanocomposite.

Polyaniline/graphene (PANI–G) nanocomposites, prepared through polymerization of aniline monomer with graphene under *in situ* conditions, were evaluated for the photodecolorization of Rose Bengal dye by Ameen et al., (2012). PANI–G nanocomposite efficiently degraded Rose Bengal dye by 56% within 180 min under visible light illumination. The photocatalytic activity of PANI–G was also much superior to that of pristine PANI. The enhancement in photocatalytic activity was ascribed to the increased e^-/h^+ pair charge separation and the formation of oxyradicals ($•O_2^-$, $•HO_2$, •OH) due to the large surface of the nanocomposite provided by graphene sheets. The same research group, recently, investigated the photocatalytic degradation of Crystal Violet by ZnO–GO nanohybrids (Ameen et al., 2013). An enormously high photodegradation of ~95% was achieved within 80 min of UV irradiation.

Herring et al., (2012) reported a simple method for the preparation of ZnO–rGO composites by using microwave irradiation of zinc acetate and GO in the presence of a mixture of oleic acid ($C_{18}H_{34}O_2$) and oleylamine ($C_{18}H_{37}N$). Microwave power density and exposure time strongly influenced the growth rate, shape, size, and morphology of ZnO nanocrystals supported on the rGO nanosheets. The as-synthesized ZnO–rGO nanocomposite was studied as a potential photocatalyst for the degradation of a diamino-triphenylmethane dye, Malachite Green. The ZnO–rGO nanocomposites were found to be more efficient than pristine ZnO nanopyramids for photodegradation of the dye under UV irradiation. The photocatalytic decomposition of Malachite Green by ZnO–rGO obeyed pseudo-first-order kinetics with high degradation rate constant. The superior performance of the nanocomposite was attributed to the efficient charge transfer of the photogenerated electrons in the conduction band of ZnO to rGO.

Pant et al., (2013a) tested Ag–ZnO/rGO composite for the degradation of Reactive Black 5 under UV illumination. The Ag–ZnO/rGO nanocomposite showed good activity for the photodecomposition of Reactive Black 5. After 200 min of UV irradiation, more than 80% of Reactive Black 5 dye was degraded. Pant et al., (2013a) explained that because of the excellent electron-accepting property of rGO, photoexcited electrons of ZnO were quickly transferred from its conduction band to the rGO sheets and anchored Ag nanoparticles. This in turn effectively suppressed the recombination of photo-generated charge carriers, leaving more charge carriers to form highly reactive species ($•O_2^-$, •OH) and promoting the degradation of Reactive Black 5.

2.2.3. Photocatalytic Degradation of Other Organic Pollutants

GSNs have also been successfully applied in the photocatalytic degradation of many different types of organic pollutants, apart from synthetic dyes. Li et al., (2012d) synthesized a chemically bonded graphene–titania/silica (G–TiO_2/SiO_2) nanocomposite with hexagonal ordered mesoporous structure. The G–TiO_2/SiO_2 composite photocatalyst decomposed 93.1% of the aqueous endocrine-disrupting chemical Atrazine after 180 min of exposure to simulated sunlight.

Recently, Li et al., (2013d) fabricated graphene/TiO_2 composites using a single-step nonionic surfactant strategy combined with the solvothermal treatment technique. The as-prepared composites were then successfully applied to degrade two persistent organic pollutants (POPs) namely, Aldicarb and Norfloxacin, under both simulated sunlight and visible light irradiation. The mineralization of aqueous POPs followed the Langmuir-Hinshelwood first-order model. The high photocatalytic performance of the composite could be attributed to the enhanced quantum efficiency, narrowed band gap, and perfect textural properties. In another study by the same research group, graphene–tourmaline–TiO_2 (G–T–TiO_2) composites, prepared using a direct sol–gel co-condensation combined with solvothermal treatment technique, displayed a higher photocatalytic activity than pure TiO_2 for degradation of Norfloxacin under simulated sunlight irradiation (Li et al., 2012e).

Ghasemi et al., (2012) reported the fabrication of CeO_2–TiO_2–graphene nancomposites by hydrothermal reaction of GO with CeO_2–TiO_2 nanoparticles in aqueous solution of ethanol. The as-prepared composite photocatalyst showed an enhanced photocatalytic activity for the degradation of herbicide, 2,4-Dichlorophenoxyacetic acid (2,4-D), under UV light. Similarly, Tang et al., (2012) also investigated the photodecomposition efficiency of Ag/rGO–TiO_2 for 2,4-D. The ternary catalyst exhibited almost 100% photocatalytic removal of 2,4-D from water under simulated solar light irradiation. The photodegradation rate toward 2,4-D over Ag/rGO–TiO_2 was 11.3 times higher than that over bare TiO_2. In another recent study, Tang et al., (2013b) prepared magnetite@SiO_2@TiO_2–graphene photocatalyst by combining sol–gel and assembling processes and found that the nanocomposite could completely remove 2,4-D from water when irradiated with simulated solar light for about 140 min. The repetitive use of magnetite@SiO_2@TiO_2–graphene was also investigated. As shown in Figure 10, the removal efficiency of 2,4-D over the photocatalyst was still as high as 97.7% after 8 successive cycles, and achieved 99.1% when the catalyst was re-treated by ultrasonication after 8 cycles. Also, after being kept aside for one year, the photocatalyst still showed a 2,4-D removal efficiency as high as 95.6% when it was treated by ultrasonic dispersion before use, while in the case without ultrasonic treatment, the removal efficiency was only 84%.

Pastrana-Martínez and colleagues (Pastrana-Martínez et al., 2012) developed rGO–TiO_2 composites by liquid phase deposition followed by post-thermal reduction. The composite was then evaluated as photocatalysts for the degradation of an important pharmaceutical water pollutant, diphenhydramine (DP). Almost complete degradation of DP was achieved under near-UV/Vis irradiation in less than 60 min. In a separate study, Pastrana-Martínez and colleagues compared the photocatalytic activity of a TiO_2 catalyst synthesized by a modified sol–gel method (ECT), TiO_2 nanoparticles surface modified with organic shell layer (m-TiO2) and a GO–TiO_2 composite for mineralization of DP (Morales-Torres et al., 2013). The authors found that under near-UV/Vis irradiation, both ECT and GO–TiO_2 were highly active photocatalysts for the degradation of DP.

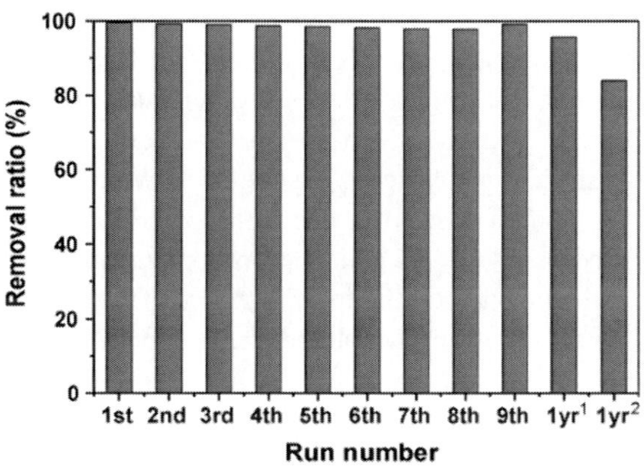

Figure 10. Degradation efficiency of 2,4-D over magnetite@SiO_2@TiO_2-graphene during 9 cycles and after being kept aside for one year (1 year). The catalyst used in the 9th cycle was treated by ultrasonic rinse. 1 yr^1 and 1 yr^2 represent the photocatalysts treated by ultrasonication or not before use, respectively. Reprinted from Tang et al., (2013b), Copyright 2013, with permission from Elsevier.

Neppolian et al., (2012) adopted an ultrasound-assisted method for synthesizing Pt–GO–TiO_2 nanocomposite and evaluated its photocatalytic efficiency for dodecylbenzenesulfonate (DBS). The Pt–GO–TiO_2 catalyst degraded DBS at a significantly higher rate than Degussa P25.

2.2.4. Photocatalytic Disinfection of Biological Pollutants

Wastewater may also contain pathogenic microorganisms (bacteria, viruses, fungi, algae, amoebas, and plankton), which pose a serious threat to public health (Kemp et al., 2013). Conventional water treatment technologies have limitations in complete decontamination of these biological pollutants without harmful disinfection by-products. Semiconductor-mediated heterogeneous photocatalysis is a promising alternative for the removal of pathogens from wastewaters since it does not generate any harmful by-products. GSNs are therefore being widely studied as photocatalyst for remediation of biological pollutants. Liu et al., (2011d) studied the comparative efficacy of GO–TiO_2 nanorod composites (GO–TiO_2 NRCs) and GO–TiO_2 nanoparticle composites (GO–TiO_2 NPCs) for the photocatalytic disinfection of water infected with *E. coli*. The GO–TiO_2 NRCs exhibited superior antibacterial activity. *E. coli* cells (at a concentration of 1.7×10^8 cfu mL^{-1}) were completely inactivated by GO–TiO_2 NRCs under solar irradiation within 120 min (Figure 11). The higher photocatalytic activity of GO–TiO_2 NRCs was due to anti-charge recombination and more {1 0 1} facets. In another separate study, Liu et al., (2012c) found that GO–Ag_3PO_4 composites, fabricated by an ion-exchange method of silver acetate (CH_3COOAg) and disodium phosphate (Na_2HPO_4) in the presence of GO sheets, showed exceptional intrinsic antibacterial activity towards *E. Coli.* Such photodecontamination efficiency was due to the synergistic effects of excellent bactericidal properties of Ag^+, good adsorption capability of GO toward bacterial cells and high photocatalytic performance of Ag_3PO_4.

Recently, Liu's group also explored the photodisinfection activities of GO–CdS composites (Gao et al., 2013). Nearly 100% of both gram-negative (*E. coli*) and gram-positive (*B. subtilis*) bacteria were killed by the GO–CdS nanocomposite within 25 min of visible light irradiation. The excellent disinfection activities was attributed to the high quantity of •OH radicals generated under visible light irradiation as confirmed by transient photocurrent measurements and radical scavenger investigations. More recently, Pant et al., (2013b) have shown that Ag–ZnO/rGO composites also have excellent intrinsic antibacterial activity towards gram-negative bacteria (*E. coli*). Moreover, the nanocomposite could be recovered easily by sedimentation for repeated use.

Cai and co-workers (Cai et al., 2011) tested the antibacterial performance of a water-soluble brilliant blue/rGO/tetradecyltriphenylphosphonium bromide composite (BB-rGO-TTP) for both gram-positive (*S. aureus*) and gram-negative (*E. coli*) bacteria, prepared by using non-covalent brilliant blue-functionalized reduced graphene oxide (BB-rGO) as the tetradecyltriphenylphosphonium bromide (TTP) carrier. The results showed that the novel BB-rGO-TTP exhibited excellent synergistic antibacterial activity, specific targeting capability, water solubility, and mild cytotoxicity, suggesting its great potential application as sprayable graphene-based antibacterial solutions. Akhavan et al., (2012) reported that graphene–WO_3 composites, fabricated by deposition of WO nanoparticles onto GO and subsequent annealing, can effectively photodegrade proteins in viruses, thereby leading to the efflux of virus RNA.

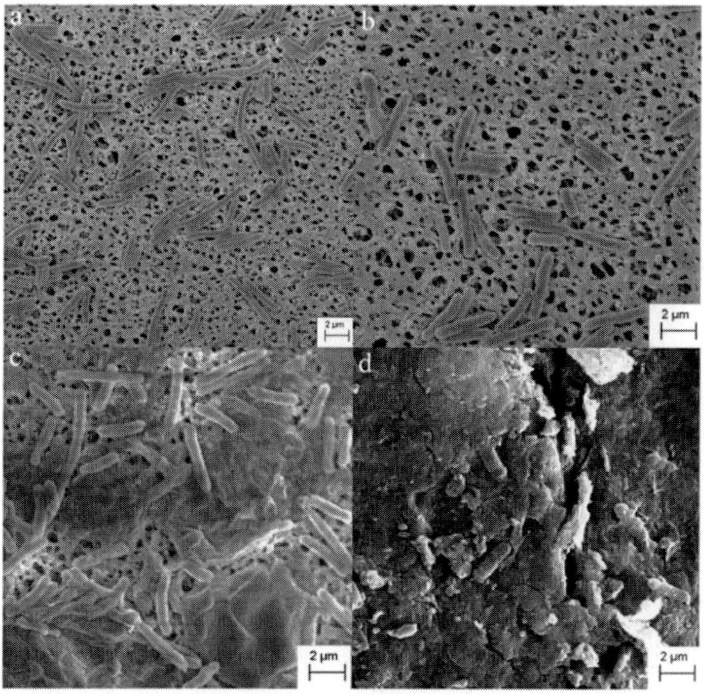

Figure 11. SEM images of E. coli (a) before solar irradiation in absence of GO–TiO_2 NRCs (b) after solar irradiation for 2 h in absence of GO–TiO_2 NRCs (c) before solar irradiation in presence of GO–TiO_2 NRCs (d) after solar irradiation for 2 h in presence of GO–TiO_2 NRCs Reprinted from Liu et al., (2011d), Copyright 2011, with permission from Elsevier.

3. Gas Separation and Purification

Air pollution, resulting from modern life styles and industrial developments, is a topic of serious environmental concern due to its detrimental impact on global atmospheric chemistry and climate as well as flora, fauna and human health. Control of air pollutants is thus necessary to provide a better and safe environment for future generations. Recently, graphene-based materials have also gained considerable popularity as adsorbent for removing various toxic and poisonous gaseous air pollutants. For example, Bandosz and co-workers have conducted and reported a number of experimental studies on adsorption of ammonia (NH_3) by graphene-based materials. They have shown that NH_3 gas can be effectively captured by modifying GO with either aluminium–zirconium polycation (Seredych and Bandosz, 2009a), Al_{13} polycations (Seredych and Bandosz, 2010a), or MnO_2 nanoparticles (Seredych and Bandosz, 2012). In addition to NH_3, other common air pollutants such as H_2S, NO_2 and SO_2 can also be effectively adsorbed using graphene-based materials. Seredych and Bandosz have found that N-doped GO, prepared by the reduction of GO using NH_3 gas, can be used to capture H_2S (Seredych and Bandosz, 2009b). The basic nature of N tends to enhance the dissociation of acidic H_2S and attracts HS^- ions via electrostatic interactions. These ions are then subsequently oxidized to elemental sulphur and a small amount of SO_2.

The same research group has also shown that iron acetato–GO composite can be used as a suitable adsorbent for removal of NO_2 from industrial waste gas (Petit and Bandosz, 2012). NO_2 is reactively adsorbed on the surface of the composite, where after it is reduced to NO. The release of NO can be counteracted by the inclusion of active iron species such as γ-FeOOH and α-Fe_2O_3, which in turn also increases the adsorption capacity.

Bandosz's group has also investigated the adsorption of SO_2 using a hydrous zirconia–graphene composite (Seredych and Bandosz, 2010b). The interactions between zirconia and graphene result in pores which increases the adsorption capacity of SO_2 under both dry and moist conditions compared to pristine GO and pure $Zr(OH)_4$, when the GO content is 5 wt%. As the GO content is increased in the nanocomposite, there is an increased pore formation due to enhanced binding between GO and $Zr(OH)_4$ leading to a higher adsorption capacity under moist conditions.

Besides being adsorbent, graphene materials can also be used as catalyst to degrade nitrogen oxides (NO_x). For instance, BiOBr-graphene composite photocatalyst have been found to have a NO removal efficiency of 40.3%

under visible light, and the photocatalytic NO removal rate constant of BiOBr-graphene was 2 times that of pure BiOBr (Ai et al., 2011b).

4. Conclusion and Future Perspectives

In this chapter, the latest research work reported in the literature on the use of graphene materials for environmental remediation has been synthesized and discussed. It is evident that graphene, and its related materials such as GO, rGO as well as their nanocomposites offer a wide range of benefits for different environmental remediation applications. Their simplicity, flexibility, large surface area, and intriguing physic-chemical properties make them attractive for environmental applications such as water purification, gas separation and storage, as discussed in this chapter. However, research in this area is still at its infancy. There are a number of challenges that have to be overcome before we can realize the full potential of graphene-based materials for practical applications. Here, we would like to highlight some of those important challenges that may be overcome from future research.

First, graphene and its derivatives are commonly synthesized by conventional methods involving the use of toxic chemicals. These methods usually result in the generation of hazardous waste and poisonous gases. Therefore, there is a need to develop environmentally benign methods to produce graphene materials by following green approaches.

Second, although graphene-based materials have been found to be effective in the field of environmental pollutant remediation, only a small range of pollutants, including organic dyes, common metal ions, and volatile gaseous pollutant, have been explored. Therefore, more studies must be conducted with respect to adsorptive removal of emerging organic pollutants, EDCs, PPCPs, volatile organics and aerosols. Also, it is extremely important to investigate the simultaneous removal of many co-existing pollutants from multicomponent systems. Such studies are essential since they provide insights into competitive removal and possible interference from other contaminants to target species.

Third, the lack of efficient and low-cost harvesting techniques and inadequate scale-up strategies pose a hindrance to the practical exploitation of graphene materials in environmental remediation. Therefore, future research work should also aim at addressing these issues.

Fourth, although several processes including ion exchange, electrostatic interactions, π-π electron donor-acceptor interactions, surface adsorption, and complexation, have been suggested to explain the mechanisms involved in the adsorption of water pollutants by graphene-based materials, mechanistic studies need to be conducted in detail to validate the proposed binding mechanism. Also, the mechanisms of pollutant catalytic degradation must be investigated and studied in detail. Finally, very little information is currently available on the toxicity and biocompatibility of graphene and related materials. Future investigations should, therefore, focus on *in vitro* and *in vivo* interactions between graphene materials and different living systems to effectively evaluate the utilization of these materials in environmental remediation.

Undoubtedly, graphene and its relatives hold a great potential for being robust materials to address the different environmental problems that the world is facing today. However, it is only when we overcome the above mentioned challenges the commercial application of graphene materials in environmental pollution control and abatement would be successfully realized.

References

Ahmad, M., Ahmed, E., Hong, Z. L., Xu, J. F., Khalid, N. R., Elhissi, A. & Ahmed, W. (2013). A facile one-step approach to synthesizing ZnO/graphene composites for enhanced degradation of Methylene Blue under visible light. *Applied Surface Science*, *274*, 273-281.

Ahmaruzzaman, M. (2008). Adsorption of phenolic compounds on low-cost adsorbents: a review. *Advances in Colloid and Interface Science*, *143*, 48-67.

Ahmed, S., Rasul, M. G., Martens, W. N., Brown, R. & Hashib, M. A. (2011). Advances in heterogeneous photocatalytic degradation of phenols and dyes in wastewater: A review. *Water Air & Soil Pollution*, *215*, 3-29.

Ai, L., Zhang, C. & Chen, Z. (2011a). Removal of methylene blue from aqueous solution by a solvothermal-synthesized graphene/magnetite composite. *Journal of Hazardous Materials*, *192*, 1515-1524.

Ai, Z., Ho, W. & Lee, S. (2011b). Efficient visible light photocatalytic removal of NO with BiOBr-graphene nanocomposites. *Journal of Physical Chemistry C*, *115*, 25330-25337.

Ai, L. & Jiang, J. (2012). Removal of methylene blue from aqueous solution with self-assembled cylindrical graphene-carbon nanotube hybrid. *Chemical Engineering Journal*, *192*, 156-163.

Akhavan, O., Choobtashani, M. & Ghaderi E. (2012). Protein degradation and RNA efflux of viruses photocatalyzed by graphene-tungsten oxide composite under visible light irradiation. *Journal of Physical Chemistry C*, *116*, 9653-9659.

Aksu, Z. (2005). Application of biosorption for the removal of organic pollutants: a review. *Process Biochemistry*, *40*, 997-1026.

Algothmi, W. M., Bandaru, N. M., Yu, Y., Shapter, J. G. & Ellis, A. V. (2013). Alginate-graphene oxide hybrid gel beads: An efficient copper adsorbent material. *Journal of Colloid and Interface Science*, *397*, 32-38.

Ameen, S., Seo, H.-K., Akhtar, M. S. & Shin, H. S. (2012). Novel graphene/polyaniline nanocomposites and its photocatalytic activity toward degradation of Rose Bengal dye. *Chemical Engineering Journal*, *210*, 220-228.

An, X., Yu, J. C., Wang, F., Li, C. & Li, Y. (2013). One-pot synthesis of In_2S_3 nanosheets/graphene compsoites with enhanced visible-light photocatalytic activity. *Applied Catalysis B: Environmental*, *129*, 80-88.

Augugliaro, V., Bellardita, M., Loddo, V., Palmisano, G., Palmisano, L. & Yurdakal, S. (2012). Overview on oxidation mechanisms of organic compounds by TiO_2 in heterogeneous photocatalysis. *Journal of Photochemistry and Photobiology C: Photochemistry Reviews*, *13*, 224-245.

Avouris, P. & Dimitrakopoulos, C. (2012). Graphene: synthesis and applications. *Materials Today*, *15*, 86-97.

Brownson, D. A. C., Kampouris, D. K. & Banks, C. E. (2011). An overview of graphene in energy production and storage applications. *Journal of Power Sources*, *196*, 4873-4885.

Bhatnagar, A. & Sillanpaa, M. (2010). Utilization of agro-industrial and municipal waste materials as potential adsorbents for water treatment —A review. *Chemical Engineering Journal*, *157*, 277-296.

Bhunia, P., Kim, G., Baik, C. & Lee, H. (2012). A strategically designed porous iron-iron oxide matrix on graphene for heavy metal adsorption. *Chemical Communications*, *48*, 9888-9890.

Bunch, J. S., Verbridge, S. S., Alden, J. S., van der Zande, A. M., Parpia, J. M., Craighead, H. G. & McEuen, P. L. (2008). Impermeable atomic membranes from graphene sheets. *Nano Letters*, *8*, 2458–2462.

Cai, X., Tan, S., Lin, M., Xie, A., Mai, W., Zhang, X., Lin, Z., Wu, T. & Liu, Y. (2011). Synergistic antibacterial brilliant blue/reduced graphene oxide/quaternary phosphonium salt composite with excellent water solubility and specific targeting capability. *Langmuir*, *27*, 7828-7835.

Chandra, V., Park, J., Chun, Y., Lee, J. W., Hwang, I. C. & Kim, K. S. (2010). Water-dispersible magnetite-reduced graphene oxide composites for arsenic removal. *ACS Nano*, *4*, 3979-3986.

Chandra, V. & Kim, K. S. (2011). Highly selective adsorption of Hg^{2+} by a polypyrrole-reduced graphene oxide composite. *Chemical Communications*, *47*, 3942-3944.

Chandra, S., Das, P., Bag, S., Bhar, R. & Pramanik, P. (2012). Mn_2O_3 decorated graphene nanosheets: An advanced material for the photocatalytic degradation of organic dyes. *Materials Science and Engineering B*, *177*, 855-861.

Chang, Y. P., Ren, C. L., Qu, J. C. & Chen, X. G. (2012a). Preparation and characterization of Fe_3O_4/graphene nanocomposite and investigation of its adsorption performance for aniline and p-choloroaniline. *Applied Surface Science*, *261*, 504-509.

Chen, Y., Chen, L., Bai, H. & Li, L. (2013a). Graphene oxide-chitosan composite hydrogels as broad-spectrum adsorbents for water purification. *Journal of Materials Chemistry A*, *1*, 1992-2001.

Chen, D., Li, Y. & Chen, Y. (2013b). Preparation of graphene and γ-Fe_2O_3 doped titanium dioxide and its photocatalytic properties. *Applied Mechanics and Materials*, 295-298, 447-451.

Cheng, J. S., Du, J. & Zhu, W. (2012). Facile synthesis of three-dimensional chitosan-graphene mesostructures for reactive black 5 removal. *Carbohydrate Polymers*, *88*, 61-67.

Cheng, H., Zeng, K. & Yu, J. (2013). Adsorption of uranium from aqueous solution by graphene oxide nanosheets supported on sepiolite. *Journal of Radioanalytical and Nuclear Chemistry*, doi: 10.1007/s10967-012-2406-6.

Compton, O. C., Kim, S., Pierre, C., Torkelson, J. M. & Nguyen, S. T. (2010). Crumpled graphene nanosheets as highly effective barrier property enhancers. *Advanced Materials*, *22*, 4759–4763.

Corcoran, E., Nellemann, C., Baker, E., Bos, R., Osborn, D. & Savelli, H. (Eds.), (2010). Sick water? The central role of wastewater management in sustainable development. A Rapid Response Assessment. United Nations Environment Programme – United Nations Human Settlements Programme (UNEP—UN-HABITAT), Arendal, Norway.

Crini, G. (2005). Recent developments in polysaccharide-based materials used as adsorbents in wastewater treatment. *Progress in Polymer Science*, *30*, 38-70.

Crini, G. (2006). Non-conventional low-cost adsorbents for dye removal: a review. *Bioresource Technology*, *97*, 1061-1085.

Crini, G. & Badot, P. M. (2008). Application of chitosan, a natural aminopolysaccharide, for dye removal from aqueous solutions by adsorption processes using batch studies: A review of recent literature. *Progress in Polymer Science*, *33*, 399-447.

Dabrowski, A. (2001). Adsorption—from theory to practice. *Advances in Colloid and Interface Science* 93, 135-224.

Delgado, L. F., Charles, P., Glucina, K. & Morlay, C. (2012). The removal of endocrine disrupting compounds, pharmaceutically activated compounds and cyanobacterial toxins during drinking water preparation using activated carbon – a review. *Science of the Total Environment*, 435-436, 509-525.

Demirbas, H. (2008). Heavy metal adsorption onto agro-based waste materials: a review. *Journal of Hazardous Materials*, *157*, 220-229.

Deng, X., Lu, L., Li, H. & Luo, F. (2010). The adsorption properties of Pb(II) and Cd(II) on functionalized graphene prepared by electrolysis method. *Journal of Hazardous Materials*, *183*, 923-930.

Dikin, D.A., Stankovich, S., Zimney, E.J., Piner, R.D., Dommett, G.H., Evmenenko, G., Nguyen, S. T. & Ruoff, R. S. (2007). Preparation and characterization of graphene oxide paper. *Nature*, *448*, 457–460.

Djilani, C., Zaghdoudi, R., Modarressi, A., Rogalski, M., Djazi, F. & Lallam, A. (2012). Elimination of organic micropollutants by adsorption on activated carbon prepared from agricultural waste. *Chemical Engineering Journal* 189-190, 203-212.

Fan, L., Luo, C., Li, X., Lu, F., Qiu, H. & Sun, M. (2012a). Fabrication of novel magnetic chitosan grafted with graphene oxide to enhance adsorption properties for methyl blue. *Journal of Hazardous Materials* 215-216, 272-279.

Fan, L., Luo, C., Sun, M., Li, X., Lu, F. & Qiu, H. (2012b). Preparation of novel magnetic chitosan/graphene oxide composite as effective adsorbents towards methylene blue. *Bioresource Technology*, *114*, 703-706.

Fan, L., Luo, C., Sun, M., Li, X. & Qiu, H. (2013a). Highly selective adsorption of lead ions by water-dispersible magnetic chitosan/graphene oxide composites. *Colloids and Surfaces B: Biointerfaces*, *103*, 523-529.

Fan, L., Luo, C., Sun, M., Qiu, H. & Li, X. (2013b). Synthesis of magnetic β-cyclodextrin-chitosan/graphene oxide as nanoadsorbent and its application in dye adsorption and removal. *Colloids and Surfaces B: Biointerfaces*, *103*, 601-607.

Farghali, A. A., Bahgat, M., El Rouby, W. M. A. & Khedr, M. H. (2013). Preparation, decoration and characterization of graphene sheets for methyl green adsorption. *Journal of Alloys and Compounds*, *555*, 193-200.

Foo, K. Y. & Hameed, B. H. (2010). Insights into the modelling of adsorption isotherm systems. *Chemical Engineering Journal*, *156*, 2-10.

Fu, Y., Sun, X. & Wang, X. (2011). $BiVO_4$—graphene catalyst and its high photocatalytic performance under visible light irradiation. *Materials Chemistry and Physics*, *131*, 325-330.

Fu, Y., Chen, H., Sun, X. & Wang, X. (2012). Combination of cobalt ferrite and graphene: High-performance and recyclable visible-light photocatalysis. *Applied Catalysis B: Environmental*, 111-112, 280-287.

Gao, Y., Li, Y., Zhang, L., Huang, H., Hu, J., Shah, S.M. & Su, X. (2012a). Adsorption and removal of tetracycline antibiotics from aqueous solution by graphene oxide. *Journal of Colloid and Interface Science*, *368*, 540-546.

Gao, Z., Liu, J., Xu, F., Wu, D., Wu, Z. & Jiang, K. (2012b). One-pot synthesis of graphene-cuprous oxide composite with enhanced photocatalytic activity. *Solid State Sciences*, *14*, 276-280.

Gao, P., Liu, J., Sun, D. D. & Ng, W. (2013). Graphene oxide—CdS composite with high photocatalytic degradation and disinfection activities under visible light irradiation. *Journal of Hazardous Materials*, 250-251, 412-420.

Geim, A. K. & Novoselov, K. S. (2007). The rise of graphene. *Nature Materials*, *6*, 183-191.

Ghasemi, S., SetayeshRahman, S., Habibi-Yangjeh, A., Hormozi-Nezhad, M. R. & Gholami, M.R. (2012). Assembly of CeO_2—TiO_2 nanoparticles prepared in room temperature ionic liquid on graphene nanosheets for photocatalytic degradation of pollutants. *Journal of Hazardous Materials*, 199-200, 170-178.

Ghasemi, S., Esfandiar, A., Setayesh, S.R., Habibi-Yangjeh, A., Irajizad, A. & Gholami, M. R. (2013). Synthesis and characterization of TiO_2—graphene nanocomposites modified with noble metals as a photocatalyst for degradation of pollutants. *Applied Catalysis A: General*, 462-463, 82-90.

Ghosh, T., Cho, K. Y., Ullah, K., Nikam, V., Park, C. Y., Meng, Z. D. & Oh, W. C. (2013). High photonic effect of organic dye degradation by CdSe—graphene—TiO_2 particles. *Journal of Industrial and Engineering Chemistry*, *19*, 797-805.

Gupta, V. K. & Suhas (2009). Application of low-cost adsorbents for dye removal—A review. *Journal of Environmental Management*, *90*, 2313-2342.

Gupta, V. K., Carrott, P. J. M. & Carrott, M. M. I. R., Suhas (2009). Low-cost adsorbents: growing approach to wastewater treatment - a review. *Critical Reviews in Environmental Science and Technology*, *39*, 783-842.

Han, L., Wang, P. & Dong, S. (2012). Progress in graphene-based photoactive nanocomposites as a promising class of photocatalyst. *Nanoscale*, *4*, 5814-5825.

Hao, L., Song, H., Zhang, L., Wan, X., Tang, Y. & Lv, Y. (2012). SiO_2/graphene composite for highly selective adsorption of Pb(II) ion. *Journal of Colloid and Interface Science*, *369*, 381-387.

He, Y. Q., Zhang, N. N. & Wang, X. D. (2011). Adsorption of graphene oxide/chitosan porous materials for metal ions. *Chinese Chemical Letters*, *22*, 859-862.

He, G. Y., Huang, J., Liu, W. F., Wang, X., Chen, H. Q. & Sun, X. Q. (2012). ZnO-Bi_2O_3/graphene oxide photocatalyst with high photocatalytic performance under visible light. *Materials Technology*, *27*, 278-283.

Herring, N. P., Almahoudi, S. H., Olson, C. R. & El-Shall, M. S. (2012). Enhanced photocatalytic activity of ZnO-graphene nanocomposites prepared by microwave synthesis. *Journal of Nanoparticle Research*, *14*, 1277.

Huang, X., Qi, X., Boey, F. & Zhang, H. (2012). Graphene-based composites. *Chemical Society Reviews*, *41*, 666-686.

Huimin, Z., Fang, S., Xinfei, F., Hongtao, Y., Dan, W. & Xie, Q. (2012). Graphene—TiO_2 composite photocatalyst with enhanced photocatalytic performance. *Chinese Journal of Catalysis*, *33*, 777-782.

Ismail, A. A., Geioushy, R. A., Bouzid, H., Al-Sayari, S. A., Al-Hajry, A. & Bahnemann, D. W. (2013). TiO_2 decoration of graphene layers for highly efficient photocatalyst: Impact of calcination at different gas atmosphere on photocatalytic efficiency. *Applied Catalysis B: Environmental*, *129*, 62-70.

Ivanovskii, A. L. (2012). Graphene-based and graphene-like materials. *Russian Chemical Reviews*, *81*, 571-605.

Jabeen, H., Chandra, V., Jung, S., Lee, J. W., Kim, K. S. & Kim, S. B. (2011). Enhanced Cr(VI) removal using iron nanoparticle decorated graphene. *Nanoscale*, *3*, 3583-3585.

Kemp, K. C., Seema, H., Saleh, M., Le, N.H., Mahesh, K., Chandra, V. & Kim, K. S. (2013). Environmental applications using graphene composites: water remediation and gas adsorption. *Nanoscale*, *5*, 3149-3171.

Khin, M. M., Nair, A. S., Babu, V. J., Murugan, R. & Ramakrishna, S. (2012). A review on nanomaterials for environmental remediation. *Energy & Environmental Science*, *5*, 8075-8109.

Kuilla, T., Bhadra, S., Yao, D., Kim, N. H., Bose, S. & Lee, J. H. (2010). Recent advances in graphene based polymer composites. *Progress in Polymer Science*, *35*, 1350-1375.

Kurniawan, T. A., Chan, G. Y. S., Lo, W. h. & Babel, S. (2006). Comparisons of low-cost adsorbents for treating wastewaters laden with heavy metals. *Science of the Total Environment*, *366*, 409-426.

Lapworth, D. J., Baran, N., Stuart, M. E. & Ward, R. S. (2012). Emerging organic contaminants in groundwater: a review of sources, fate and occurrence. *Environmental Pollution*, *163*, 287-303.

Leary, R. & Westwood, A., (2011). Carbonaceous nanomaterials for the enhancement of TiO_2 photocatalysis. *Carbon*, *49*, 741-772.

Lee, Y. C. & Yang, J. W. (2012). Self-assembled flower-like TiO_2 on exfoliated graphite oxide for heavy metal removal. *Journal of Industrial and Engineering Chemistry*, *18*, 1178-1185.

Lee, E., Hong, J. Y., Kang, H. & Jang, J. (2012). Synthesis of TiO_2 nanorod-decorated graphene sheets and their highly efficient photocatalytic activities under visible-light irradiation. *Journal of Hazardous Materials*, 219-220, 13-18.

Leng, Y., Guo, W., Su, S., Yi, C. & Xing, L. (2012). Removal of antimony(III) from aqueous solution by graphene as an adsorbent. *Chemical Engineering Journal*, 211-212, 406-411.

Li, Y.H., Liu, T., Du, Q., Sun, J., Xia, Y., Wang, Z., Zhang, W., Wang, K., Zhu, H. & Wu, D. (2011a). Adsorption of Cationic Red X-GRL from aqueous solutions by graphene: Equilibrium, kinetics and thermodynamics study. *Chemical and Biochemical Engineering Quarterly*, *25*, 483-491.

Li, N., Zheng, M., Chang, X., Ji, G., Lu, H., Xue, L., Pan, L. & Cao, J. (2011b). Preparation of magnetic $CoFe_2O_4$-functionalized graphene sheets via a facile hydrothermal method and their adsorption properties. *Journal of Solid State Chemistry*, *184*, 953-958.

Li, Z., Chen, F., Yuan, L., Liu, Y., Zhao, Y., Chai, Z. & Shi, W. (2012a). Uranium(VI) adsorption on graphene oxide nanosheets from aqueous solutions. *Chemical Engineering Journal*, *210*, 539-546.

Li, S., Lu, X., Xue, Y., Lei, J., Zheng, T. & Wang, C. (2012b). Fabrication of polypyrrole/graphene oxide composite nanosheets and their applications for Cr(VI) removal in aqueous solution. *PLoS One*, *7*, e43328.

Li, Y., Du, Q., Liu, T., Sun, J., Jiao, Y., Xia, Y., Xia, L., Wang, Z., Zhang, W., Wang, K., Zhu, H. & Wu, D. (2012c). Equilibrium, kinetic and thermodynamic studies on the adsorption of phenol onto graphene. *Materials Research Bulletin*, *47*, 1898-1904.

Li, K., Huang, Y., Yan, L., Dai, Y., Xue, K., Guo, H., Huang, Z. & Xiong, J. (2012d). Simulated sunlight photodegradation of aqueous atrazine and rhodamine B catalysed by the ordered mesoporous graphene-titania/silica composite materials. *Catalysis Communications*, *18*, 16-20.

Li, K., Chen, T., Yan, L., Dai, Y., Huang, Z., Guo, H., Jiang, L., Gao, X., Xiong, J. & Song, D. (2012e). Synthesis of mesoporous graphene and tourmaline co-doped titania composites and their photocatalytic activity towards organic pollutant degradation and eutrophic water treatment. *Catalysis Communications*, *28*, 196-201.

Li, L., Fan, L., Sun, M., Qiu, H., Li, X., Duan, H. & Luo, C. (2013a). Adsorbent for chromium removal based on graphene oxide functionalized with magnetic cyclodextrin-chitosan. *Colloids and Surfaces B: Biointerfaces*, *107*, 76-83.

Li, X., Zhang, T., Gu, S., Kang, S. Z., Li, G. & Mu, J. (2013b). Reduced graphene oxide/potassium niobate composite nanoscrolls with enhanced photocatalytic activity for dye degradation. *Separation and Purification Technology*, *108*, 139-142.

Li, J., Zhou, S. l., Hong, G. B. & Chang, C. T. (2013c). Hydrothermal preparation of P25-graphene composite with enhanced adsorption and photocatalytic degradation of dyes. *Chemical Engineering Journal*, *219*, 486-491.

Li, K., Xiong, J., Chen, T., Yan, L., Dai, Y., Song, D., Lv, Y. & Zeng, Z., (2013d). Preparation of graphene/TiO_2 composites by non-ionic surfactant strategy and their simulated sunlight and visible light photoactivity towards representative aqueous POPs degradation. *Journal of Hazardous Materials*, 250-251, 19-28.

Liu, X., Pan, L., Lv, T., Zhu, G., Lu, T., Sun, Z. & Sun, C. (2011a). Microwave-assisted synthesis of TiO_2-reduced graphene oxide composites for the photocatalytic reduction of Cr(VI). *RSC Advances*, 1, 1245-1249.

Liu, X., Pan, L., Lv, T., Lu, T., Zhu, G., Sun, Z. & Sun, C. (2011b). Microwave-assisted synthesis of ZnO-graphene composite for photocatalytic reduction of Cr(VI). *Catalysis Science & Technology, 1*, 1189-1193.

Liu, X., Pan, L., Lv, T., Zhu, G., Sun, Z. & Sun, C. (2011c). Microwave-assisted synthesis of CdS-reduced graphene oxide composites for photocatalytic reduction of Cr(VI). *Chemical Communications, 47*, 11984-11986.

Liu, J., Liu, L., Bai, H., Wang, Y. & Sun, D. D. (2011d). Gram-scale production of graphene oxide—TiO_2 nanorod composites: Towards high-activity photocatalytic materials. *Applied Catalysis B: Environmental, 106*, 76-82.

Liu, T., Li, Y., Du, Q., Sun, J., Jio, Y., Yang, G., Wang, Z., Xia, Y., Zhang, W., Wang, K., Zhu, H. & Wu, D. (2012a). Adsorption of methylene blue from aqueous solution by graphene. *Colloids and Surfaces B: Biointerfaces, 90*, 197-203.

Liu, Z., Xu, W., Fang, J., Xu, X., Wu, S., Zhu, X. & Chen, Z. (2012b). Decoration of BiOI quantum size nanoparticles with reduced graphene oxide in enhanced visible-light-driven photocatalytic studies. *Applied Surface Science, 259*, 441-447.

Liu, L., Liu, J. & Sun, D. D. (2012c). Graphene oxide enwrapped Ag_3PO_4 composite: towards a highly efficient and stable visible-light-induced photocatalyst for water purification. *Catalysis Science & Technology, 2*, 2525-2532.

Liu, H., Cao, W. R., Su, Y., Chen, Z. & Wang, Y. (2013). Bismuth oxyiodide-graphene nanocomposites with high visible light photocatalytic activity. *Journal of Colloid and Interface Science, 398*, 161-167.

Luo, X., Wang, C., Wang, L., Deng, F., Luo, S., Tu, X. & Au, C. (2013). Nanocomposites of graphene oxide-hydrated zirconium oxide for simultaneous removal of As(III) and As(V) from water. *Chemical Engineering Journal, 220*, 98-106.

Ma, H., Shen, J., Shi, M., Lu, X., Li, Z., Yong, Y., Li, N. & Ye, M. (2012). Significant enhanced performance for Rhodamine B, phenol and Cr(VI) removal by Bi_2WO_6 nancomposites via reduced graphene oxide modification. *Applied Catalysis B: Environmental*, 121-122, 198-205.

Madadrang, C. J., Kim, H. Y., Gao, G., Wang, N., Zhu, J., Feng, H., Gorring, M., Kasner, M. L. & Hou, S. (2012). Adsorption behaviour of EDTA-graphene oxide for Pb(II) removal. *ACS Applied Materials and Interfaces, 4*, 1186-1193.

Malaviya, P. & Singh, A. (2011). Physicochemical technologies for remediation of chromium-containing waters and wastewaters. *Critical Reviews in Environmental Science and Technology*, *41*, 1111-1172.

Mi, X., Huang, G., Xie, W., Wang, W., Liu, Y. & Gao, J. (2012). Preparation of graphene oxide aerogel and its adsorption for Cu^{2+} ions. *Carbon*, *50*, 4856-4864.

Min, Y., Zhang, F.-J., Zhao, W., Zheng, F., Chen, Y. & Zhang, Y. (2012). Hydrothermal synthesis of nanosized bismuth niobate and enhanced photocatalytic activity by coupling graphene sheets. *Chemical Engineering Journal*, *209*, 215-222.

Morales-Torres, S., Pastrana-Martínez, L. M., Figueiredo, J. L., Faria, J. L. & Silva, A. M. T. (2013). Graphene oxide-P25 photocatalysts for degradation of diphenhydramine pharmaceutical and methyl orange dye. *Applied Surface Science*, *275*, 361-268.

Neppolian, B., Bruno, A., Bianchi, C. L. & Ashokkumar, M. (2012). Graphene oxide based Pt-TiO_2 photocatalyst: Ultrasound assisted synthesis, characterization and catalytic efficiency. *Ultrasonics Sonochemistry*, *19*, 9-5.

Nguyen-Phan, T. D., Pham, V. H., Shin, E. W., Pham, H. D., Kim, S., Chung, J. S., Kim, E. J. & Hur, S. H. (2011). The role of graphene oxide content on the adsorption-enhanced photocatalysis of titanium dioxide/graphene oxide composites. *Chemical Engineering Journal*, *170*, 226-232.

Novoselov, K. S., Fal'ko, V. I., Colombo, L., Gellert, P. R., Schwab, M. G. & Kim, K. (2012). *Nature*, *490*, 192-200.

Paek, S. M., Yoo, E. & Honma, I. (2009). Enhanced cyclic performance and lithium storage capacity of SnO_2/graphene nanoporous electrodes with three-dimensionally delaminated flexible structure. *Nano Letters*, *9*, 72–75.

Pan, S. & Liu, X. (2012). CdS-Graphene nanocomposite: synthesis, adsorption kinetics and high photocatalytic performance under visible light irradiation. *New Journal of Chemistry*, *36*, 1781-1787.

Pant, H. R., Pant, B., Kim, H. J., Amarjargal, A., Park, C. H., Tijing, L. D., Kim, E. K. & Kim, C. S. (2013a). A green and facile one-pot synthesis of Ag-ZnO/RGO nanocomposite with effective photocatalytic activity for removal of organic pollutants. *Ceramics International*, *39*, 5083-5091.

Pant, H. R., Park, C. H., Pokharel, P., Tijing, L. D., Lee, D. S. & Kim, C. S. (2013b). ZnO micro-flowers assembled on reduced graphene sheets with high photocatalytic activity for removal of pollutants. *Powder Technology*, *235*, 853-858.

Pastrana-Martínez, L. M., Morales-Torres, S., Likodimos, V., Figueiredo, J. L., Faria, J. L., Falaras, P. & Silva, A. M. T. (2012). Advanced nanostructured photocatalysts based on reduced graphene oxide-TiO_2 composites for degradation of diphenhydramine pharmaceutical and methyl orange dye. *Applied Catalysis B: Environmental*, 123-124, 241-256.

Pavagadhi S., Tang, A. L. L., Sathishkumar, M., Loh, K. P. & Balasubramanian, R. (2013). Removal of microcystin-LR and microcystin-RR by graphene oxide: Adsorption and kinetic experiments. *Water Research*, *47*, 4621-4629.

Pawar, R. C., Cho, D. & Lee, C. S. (2013). Fabrication of nanocomposite photocatalysts from zinc oxide nanostructures and reduced graphene oxide. *Current Applied Physics*, *13*, S50-S57.

Pei, Z., Li, L., Sun, L., Zhang, S., Shan, X. q., Yang, S. & Wen, B. (2013). Adsorption characteristics of 1,2,4-trichlorobenzene, 2,4,6-trichlorophenol, 2-napthol and napthalene on graphene and graphene oxide. *Carbon*, *51*, 156-163.

Peng, F., Luo, T., Qiu, L. & Yuan, Y. (2013). An easy method to synthesize graphene oxide-FeOOH composites and their potential application in water purification. *Materials Research Bulletin*, *48*, 2180-2185.

Petit, C. & Bandosz, T. J. (2012). Exploring the coordination chemistry of MOF-graphite oxide composites and their applications as adsorbents. *Dalton Transactions*, *41*, 4027-4035.

Ramesha, G. K., Kumara, A. V., Muralidhara, H. B. & Sampath, S. (2011). Graphene and graphene oxide as effective adsorbents toward anionic and cationic dyes. *Journal of Colloid and Interface Science*, *361*, 270-277.

Ren, Y., Yan, N., Wen, Q., Fan, Z., Wei, T., Zhang, M., Ma, J. (2011). Graphene/δ-MnO_2 composite as adsorbent for the removal of nickel ions from wastewater. *Chemical Engineering Journal*, *175*, 1-7.

Ren, Y., Yan, N., Feng, J., Ma, J., Wen, Q., Li, N. & Dong, Q. (2012). Adsorption mechanism of copper and lead ions onto graphene nanosheet/δ-MnO_2. *Materials Chemistry and Physics*, *136*, 538-544.

Ruoff, R. (2008). Graphene: calling all chemists. *Nature Nanotechnology* 3, 10–11.

Ryoo, S. R., Kim, Y. K., Kim, M. H. & Min, D. H. (2010). Behaviors of NIH-3T3 fibroblasts on graphene/carbon nanotubes: proliferation, focal adhesion, and gene transfection studies. *ACS Nano*, *4*, 6587–6598.

Sanchez, V. C., Jachak, A., Hurt, R. H. & Kane, A. B. (2011). Biological interactions of graphene-family nanomaterials: An interdisciplinary review. *Chemical Research in Toxicology*, *25*, 15-24.

Sen Gupta, S., Sreeprasad, T. S., Maliyekkal, S. M., Das, S. K. & Pradeep, T. (2012). Graphene from sugar and its application in water purification. *ACS Applied Materials and Interfaces*, *4*, 4156-4163.

Seredych, M. & Bandosz, T. J. (2009a). Graphite oxide/AlZr polycation composites: Surface characterization and performance as adsorbents of ammonia. *Materials Chemistry and Physics*, *117*, 99-106.

Seredych, M. & Bandosz, T. J. (2009b). Adsorption of hydrogen sulphide on graphite derived materials modified by incorporation of nitrogen. *Materials Chemistry and Physics*, *113*, 946-952.

Seredych, M. & Bandosz, T. J. (2010a). Adsorption of ammonia on graphite oxide/Al_{13} composite. *Colloids and Surfaces A: Physicochemical and Engineering Aspects*, *353*, 30-36.

Seredych, M. & Bandosz, T. J. (2010b). Effects of surface features on adsorption of SO_2 on graphite oxide/$Zr(OH)_4$ composites. *Journal of Physical Chemistry C*, *114*, 14552-14560.

Seredych, M. & Bandosz, T. J. (2012). Manganese oxide and graphite oxide/MnO_2 composite as reactive adsorbents of ammonia at ambient conditions. *Microporous and Mesoporous Matertials*, *150*, 55-63.

Shannon, M. A., Bohn, P. W., Elimelech, M., Georgiadis, J. G., Marinas, B. J. & Mayes, A. M. (2008). Science and technology for water purification in the coming decades. *Nature*, *452*, 301-310.

Sharma, P. & Das, M. R. (2013). Removal of a cationic dye from aqueous solution using graphene oxide nanosheets: investigation of adsorption parameters. *Journal of Chemical & Engineering Data*, *58*, 151-158.

Shi, M., Shen, J., Ma, H., Li, Z., Lu, X., Li, N. & Ye, M. (2012). Preparation of graphene–TiO_2 composite by hydrothermal method from peroxotitanium acid and its photocatalytic properties. *Colloids and Surfaces A: Physicochemical and Engineering Aspects*, *405*, 30-37.

Singh, V., Joung, D., Zhai, L., Das, S., Khondaker, S. I. & Seal, S. (2011). Graphene based materials: past, present and future. *Progress in Materials Science*, *56*, 1178-1271.

Sitko, R., Turek, E., Zawisza, B., Malicka, E., Talik, E., Heimann, J. & Gagor, A. (2013). Adsorption of divalent metal ions from aqueous solutions using graphene oxide. *Dalton Transactions*, *42*, 5682-5689.

Song, S., Gao, W., Wang, X., Li, X., Liu, D., Xing, Y. & Zhang, H. (2012). Microwave-assisted synthesis of BiOBr/graphene nanocomposites and their enhanced photocatalytic activity. *Dalton Transactions*, *41*, 10472-10476.

Sreeprasad, T. S., Maliyekkal, S. M., Lisha, K. P. & Pradeep, T. (2011). Reduced graphene oxide-metal/metal oxide composites: facile synthesis and application in water purification. *Journal of Hazardous Materials*, *186*, 921-931.

Sreeprasad, T. S., Sen Gupta, S., Maliyekkal, S. M. & Pradeep, T. (2013). Immobilized graphene-based composite from asphalt: facile synthesis and application in water purification. *Journal of Hazardous Materials*, 246-247, 213-220.

Stankovich, S., Dikin, D. A., Dommett, G. H. B., Kohlhass, K. M., Zimney, E. J., Stach, E. A., Piner, R. D., Nguyen, S. T. & Ruoff, R. S. (2006). Graphene-based composite materials. *Nature*, *442*, 282-286.

Su, F. Y., You, C., He, Y. B., Lv, W., Cui, W., Jin, F., Li, B., Yang, Q. H. & Kang, F. (2010). Flexible and planar graphene conductive additives for lithium-ion batteries. *Journal of Materials Chemistry*, *20*, 9644–9650.

Sud, D. & Kaur, P. (2012). Heterogeneous photocatalytic degradation of selected organophosphate pesticides: A review. *Critical Reviews in Environmental Science and Technology*, *42*, 2365-2407.

Sui, Z., Meng, Q., Zhang, X., Ma, R. & Cao, B. (2012). Green synthesis of carbon nanotube-graphene hybrid aerogels and their use as versatile agents for water purification. *Journal of Materials Chemistry*, *22*, 8767-8771.

Sun, H., Cao, L., Lu, L., 2011. Magnetite/reduced graphene oxide nanocomposites: One step solvothermal synthesis and uses as a novel platform for removal of dye pollutants. *Nano Research*, *4*, 550-562.

Sun, L., Yu, H. & Fugetsu, B. (2012). Graphene oxide adsorption enhanced by in situ reduction with sodium hydrosulfite to remove acridine orange from aqueous solution. *Journal of Hazardous Materials*, 203-204, 101-110.

Tang, Y., Luo, S., Teng, Y., Liu, C., Xu, X., Zhang, X. & Chen, L. (2012). Efficient removal of herbicide 2,4-dichlorophenoxyacetic acid from water using Ag/reduced graphene oxide co-decorated TiO_2 nanotube arrays. *Journal of Hazardous Materials*, 241-242, 323-330.

Tang, Y., Guo, H., Xiao, L., Yu, S., Gao, N. & Wang, Y. (2013a). Synthesis of reduced graphene oxide/magnetite composites and investigation of their adsorption performance of fluoroquinolone. *Colloids and Surfaces A: Physicochemical and Engineering Aspects*, *424*, 74-80.

Tang, Y., Zhang, G., Liu, C., Luo, S., Xu, X., Chen, L. & Wang, B. (2013b). Magnetic TiO_2-graphene composite as a high-performance and recyclable platform for efficient photocatalytic removal of herbicides from water. *Journal of Hazardous Materials*, 252-253, 115-122.

Tao, W., Chang, J., Wu, D., Gao, Z., Duan, X., Xu, F. & Jiang, K. (2013). Solvothermal synthesis of graphene–Sb_2S_3 composite and the degradation activity under visible light. *Materials Research Bulletin*, *48*, 538-543.

Tu, X., Luo, S., Chen, G. & Li, J. (2012). One-pot synthesis, characterization, and enhanced photocatalytic activity of a BiOBr-graphene composite. *Chemistry – A European Journal*, *18*, 14359-14366.

Tiwari, J. N., Mahesh, K., Le, N. H., Kemp, K. C., Timilsina, R., Tiwari, R. N. & Kim, K. S. (2013). Reduced graphene-oxide based hydrogels for the efficient capture of dye pollutants from aqueous solutions. *Carbon*, *56*, 173-182.

Wang, S., Ang, P. K., Wang, Z., Tang, A. L., Thong, J. T. & Loh, K. P. (2010). High mobility, printable, and solution-processed graphene electronics. *Nano Letters*, *10*, 92–98.

Wang, C., Feng, C., Gao, Y., Ma, X., Wu, Q. & Wang, Z. (2011). Preparation of a graphene-based magnetic nanocomposite for the removal of an organic dye from aqueous solution. *Chemical Engineering Journal*, *173*, 92-97.

Wang, J. L. & Xu, L. J. (2012). Advanced oxidation processes for wastewater treatment: Formation of hydroxyl radical and application. *Critical Reviews in Environmental Science and Technology*, *42*, 251-325.

Wang, J., Tang, B., Tsuzuki, T., Liu, Q., Hou, X. & Sun, L. (2012a). Synthesis, characterization and adsorption properties of superparamagnetic polystyrene/Fe_3O_4/graphene oxide. *Chemical Engineering Journal*, 204-206, 258-263.

Wang, J., Tsuzuki, T., Tang, B., Hou, X., Sun, L. & Wang, X. (2012b). Reduced graphene oxide/ZnO composite: reusable adsorbent for pollutant management. *ACS Applied Materials and Interfaces*, *4*, 3084-3090.

Wang, P., Ao, Y., Wang, C., Hou, J. & Qian, J. (2012c). A one-pot method for the preparation of graphene—Bi_2MoO_6 hybrid photocatalyts that are responsive to visible-light and have excellent photocatalytic activity in the degradation of organic pollutants. *Carbon*, *50*, 5256-5264.

Wang, H., Yuan, X., Wu, Y., Huang, H., Peng, X., Zeng, G., Zhong, G., Liang, J. & Ren, M. (2013a). Graphene-based materials: Fabrication, characterization and application for the decontamination of wastewater and wastegas and hydrogen storage/generation. *Advances in Colloid and Interface Science*, 195-196, 19-40.

Wang, H., Yuan, X., Wu, Y., Huang, H., Zeng, G., Liu, Y., Wang, X., Lin, N. & Qi, Y. (2013b). Adsorption characteristics and behaviors of graphene oxide for Zn(II) removal from aqueous solution. *Applied Surface Science*, *279*, 432-440.

Wang, S., Wei, J., Lv, S.,Guo, Z. & Jiang, F. (2013c). Removal of organic dyes in environmental water magnetic-sulfonic graphene nanocomposite. *Clean Soil Air Water* doi. 10.1002/clen.201200460.

Wang, J., Wang, P., Cao, Y., Chen, J., Li, W., Shao, Y., Zheng, Y. & Li, D. (2013d). A high efficient photocatalyst Ag_3VO_4/TiO_2/graphene nanocomposite with wide spectral response. *Applied Catalysis B: Environmental*, 136-137, 94-102.

Wei, A., Xiong, L., Sun, L., Liu, Y., Li, W., Lai, W., Liu, X., Wang, L., Huang, W. & Dong, X. (2013). One-step electrochemical synthesis of a graphene-ZnO hybrid for improved photocatalytic activity. *Materials Research Bulletin*, *48*, 2855-2860.

Wu, T., Cai, X., Tan, S., Li, H., Liu, J. & Yang, W. (2011). Adsorption characteristics of acrylonitrile, p-toluenesulfonic acid, 1-napthalenesulfonic acid and methyl blue on graphene in aqueous solutions. *Chemical Engineering Journal*, *173*, 144-149.

Wu, Z. S., Zhou, G., Yin, L. C., Ren, W., Li, F. & Cheng, H. M. (2012). Graphene/metal oxide composite electrode materials for energy storage. *Nano Energy*, *1*, 107-131.

Wu, Y., Luo, H., Wang, H., Wang, C., Zhang, J. & Zhang, Z. (2013a). Adsorption of hexavalent chromium from aqueous solutions by graphene modified with cetyltrimethylammonium bromide. *Journal of Colloid and Interface Science*, *394*, 183-191.

Wu, W., Yang, Y., Zhou, H., Ye, T., Huang, Z., Liu, R. & Kuang, Y. (2013b). Highly efficient removal of Cu(II) from aqueous solution by using graphene oxide. *Water, Air, & Soil Pollution*, *224*, 1372-1379.

Wu, Q., Feng, C., Wang, C. & Wang, Z. (2013c). A facile one-pot solvothermal method to produce supramagnetic graphene-Fe_3O_4 nanocomposite and its application in the removal of dye from aqueous solution. *Colloids and Surfaces B: Biointerfaces*, *101*, 210-214.

Xu, T., Zhang, L., Cheng, H. & Zhu, Y. (2011). Significantly enhanced photocatalytic performance of ZnO via graphene hybridization and the mechanism study. *Applied Catalysis B: Environmental*, *101*, 382-387.

Xu, J., Wang, L., Zhu, Y. (2012). Decontamination of bisphenol A from aqueous solution by graphene adsorption. *Langmuir* 28, 8418-8425.

Xu, J., Ao, Y. & Chen, M. (2013). A simple method for the preparation of Bi_2WO_6-reduced graphene oxide with enhanced photocatalytic activity under visible light irradiation. *Materials Letters*, *92*, 126-128.

Yang, S. T., Chen, S., Chang, Y., Cao, A., Liu, Y. & Wang, H. (2011). Removal of methylene blue from aqueous solution by graphene oxide. *Journal of Colloid and Interface Science*, *359*, 24-29.

Yang, X., Chen, C., Li, J., Zhao, G., Ren, X. & Wang, X. (2012). Graphene oxide-iron oxide and reduced graphene oxide-iron oxide hybrid materials for the removal of organic and inorganic pollutants. *RSC Advances*, *2*, 8821-8826.

Yang, X., Li, J., Wen, T., Ren, X., Huang, Y. & Wang, X. (2013). Adsorption of naphthalene and its derivatives on magnetic graphene composites and the mechanism investigation. *Colloids and Surfaces A: Physicochemical and Engineering Aspects*, *422*, 118-125.

Yao, Y., Miao, S., Liu, S., Ma, L. P., Sun, H. & Wang, S. (2012a). Synthesis, characterization, and adsorption properties of magnetic Fe_3O_4@graphene nanocomposite. *Chemical Engineering Journal*, *184*, 326-332.

Yao, Y., Miao, S., Yu, S., Ma, L.P., Sun, H. & Wang, S. (2012b). Fabrication of Fe_3O_4/SiO_2 core/shell nanoparticles attached to graphene oxide and its use as an adsorbent. *Journal of Colloid and Interface Science*, *379*, 20-26.

Zeng, S., Gan, N., Weideman-Mera, R., Cao, Y., Li, T. & Sang, W. (2013). Enrichment of polychlorinated biphenyl 28 from aqueous solutions using Fe_3O_4 grafted graphene oxide. *Chemical Engineering Journal*, *218*, 108-115.

Zhang, L., Xia, J., Zhao, Q., Liu, L. & Zhang, Z. (2010a). Functional graphene oxide as a nanocarrier for controlled loading and targeted delivery of mixed anticancer drugs. *Small*, *6*, 537–544.

Zhang, K., Dwivedi, V., Chi, C. & Wu, J. (2010b). Graphene oxide/ferric hydroxide composites for efficient arsenate removal from drinking water. *Journal of Hazardous Materials*, *182*, 162-168.

Zhang, N., Qiu, H., Si, Y., Wang, W. & Gao, J. (2011). Fabrication of highly porous biodegradable monoliths strengthened by graphene oxide and their application of metal ions. *Carbon*, *49*, 827-837.

Zhang, N., Zhang, Y. & Xu, Y. J. (2012a). Recent progress on graphene-based photocatalysts: current status and future perspectives. *Nanoscale*, *4*, 5792-5813.

Zhang, K., Kemp, K. C. & Chandra, V. (2012b). Homogeneous anchoring of TiO_2 nanoparticles on graphene sheets for waste water treatment. *Materials Letters*, *81*, 127-130.

Zhang, Q., Tian, C., Wu, A., Tan, T., Sun, L., Wang, L. & Fu, H. (2012c). A facile one-pot route for the controllable growth of small sized and well-dispersed ZnO particles on GO-derived graphene. *Journal of Materials Chemistry*, *22*, 11778-11784.

Zhang, Y., Tang, Y., Li, S., Yu, S. (2013). Sorption and removal of tetrabromobisphenol A from solution by graphene oxide. *Chemical Engineering Journal*, *222*, 94-100.

Zhao, G., Ren, X., Gao, X., Tan, X., Li, J., Chen, C., Huang, Y. & Wang, X. (2011a). Removal of Pb(II) ions from aqueous solution on few-layered graphene oxide nanosheets. *Dalton Transactions*, *40*, 10945-10952.

Zhao, G., Li, J., Ren, X., Chen, C. & Wang, X. (2011b). Few-layered graphene oxide nanosheets as superior sorbents for heavy metal ion pollution management. *Environmental Science and Technology*, *45*, 10454-10462.

Zhao, G., Li, J. & Wang, X. (2011c). Kinetic and thermodynamic study of 1-napthol adsorption from aqueous solution to sulfonated graphene nanosheets. *Chemical Engineering Journal*, *173*, 185-190.

Zhao, J., Ren, W. & Cheng, H.-M. (2012a). Graphene sponge for efficient and repeatable adsorption and desorption of water contaminants. *Journal of Materials Chemistry*, *22*, 20197-20202.

Zhao, D., Sheng, G., Chen, C. & Wang, X. (2012b). Enhanced photocatalytic degradation of methylene blue under visible irradiation on graphene@TiO_2 dyade structure. *Applied Catalysis B: Environmental*, 111-112, 303-308.

Zhou, K., Zhu, Y., Yang, X., Jiang, X. & Li, C. (2011a). Preparation of graphene—TiO_2 composites with enhanced photocatalytic activity. *New Journal of Chemistry*, *35*, 353-359.

Zhou, F., Shi, R., Zhu, Y., 2011b. Significant enhancement of the visible photocatalytic degradation performances of γ-Bi_2MoO_6 nanoplate by graphene hybridization. *Journal of Molecular Catalysis A: Chemical*, *340*, 77-82.

Zhou, M., Yan, J., Cui, P. (2012). Synthesis and enhanced photocatalytic performance of WO_3 nanorods@graphene nanocomposites. *Materials Letters*, *89*, 258-261.

Zhu, Y., Murali, S., Cai, W., Li, X., Suk, J.W., Potts, J.R., Ruoff, R.S. (2010). Graphene and graphene oxide: synthesis, properties and applications. *Advanced Materials*, *22*, 3906-3924.

Zhu, J., Wei, S., Gu, H., Rapole, S.B., Wang, Q., Luo, Z., Haldolaarachchige, N., Young, D.P. & Guo, Z. (2011). One-pot synthesis of magnetic graphene nanocomposites decorated with core@double-shell nanoparticles for fast chromium removal. *Environmental Science & Technology*, *46*, 977-985.

Zong, P., Wang, S., Zhao, Y., Wang, H., Pan, H. & He, C. (2013). Synthesis and application of magnetic graphene/iron oxides composite for the removal of U(VI) from aqueous solutions. *Chemical Engineering Journal*, *220*, 45-52.

In: Graphene
Editor: Bruce T. Edwards
ISBN: 978-1-62948-795-3

Chapter 7

Configurations of Structural Defects in Graphene and Their Effects on Its Transport Properties

***T. M. Radchenko*[1,*], *V. A. Tatarenko*[1,†], *I. Yu. Sagalianov*[2,‡] and *Yu. I. Prylutskyy*[2,§]**

[1] G. V. Kurdyumov Institute for Metal Physics, N.A.S.U., Kyiv, Ukraine

[2] Taras Shevchenko National University of Kyiv, Kyiv, Ukraine

Abstract

The chapter combines analytical (statistical-thermodynamic and kinetic) with numerical (Kubo–Greenwood-formalism-based) approaches used to ascertain an influence of the configurations of point (impurities, vacancies) and line (grain boundaries, atomic steps) defects on the charge transport in graphene. Possible substitutional and interstitial graphene-based superstructures are predicted and described. The arrangements of dopants

*E-mail address: tarad@imp.kiev.ua
†E-mail address: tatar@imp.kiev.ua
‡E-mail address: isagal@ukr.net
§E-mail address: prylut@ukr.net

over sites or interstices related with interatomic-interaction energies governing the configurations of impurities. Depending on whether the interatomic interactions are short- or long-range, the low-temperature stability diagrams in terms of interaction-energy parameters are obtained. The dominance of intersublattice interactions in competition with intrasublattice ones results in a nonmonotony of ordering-process kinetics. Spatial correlations of impurities do not affect the electronic conductivity of graphene for the most important experimentally-relevant cases of point defects, neutral adatoms and screened charged impurities, while atomic ordering can give rise in the conductivity up to tens times for weak and strong short-range potentials. There is no ordering effect manifestation for long-range potentials. The anisotropy of the conductivity along and across the line defects is revealed and gives rise in the conductivity of graphene with correlated line defects as compared with the case of random ones. Simultaneously correlated (and/or ordered) point and line defects in graphene can give rise in the conductivity up to hundreds times vs. their random distribution. On an example of different B or N doping configurations in graphene, results from the Kubo–Greenwood approach are compared with those obtained from DFT method.

PACS: 61.48.Gh, 61.72.Cc, 64.60.Cn, 64.70.Nd, 68.65.-k, 72.80.Vp, 73.63.-b, 81.05.ue

Keywords: Interatomic correlation, atomic ordering, electron scattering, conductivity

Introduction

Pure and structurally perfect graphene has shown outstanding electronic phenomena such as ballistic electron propagation with extremely high carrier mobilities [1] or the quantum Hall effect at room temperature [2]. However, the absence a band-gap in pristine graphene makes its current–voltage characteristic symmetrical with respect to the zero-voltage point and thereby does not allow switching of graphene-based transistors with a high on–off ratio. There are different ways to induce a band-gap in graphene, particularly by the introduction of impurities (point defects) [3].

Generally, different types of defects are always present in crystals due to the imperfection of material production processes. Such lattice imperfections strongly affect different properties in solids. Defect in bulk crystals are investigated for many years, while graphene-based materials are considered only recently. Investigation of its transport properties and understanding factors that affect its conductivity represent one of the central directions of graphene research [4, 5, 6]. This is motivated by both fundamental interest to graphene's transport properties as well as by potential applications of this novel material for electronics. It is commonly recognized that the major factors determining the electron mobility in graphene are long-range charged impurities trapped on the substrate and strong resonant short-range scatterers due to adatoms covalently bound to graphene [5]. The nature of impurity atoms, acting as the scatterers, is directly reflected in the dependence of the conductivity on the electron density, $\sigma = \sigma(n_e)$, and therefore investigation of this function constitutes the focus of experimental and theoretical research [5, 6]. Dopant atoms change the band structure strongly dependent on atomic order and, consequently, provide a tool to govern and even to control electrical conductivity of the graphene-based materials.

Defects, playing role of disorder, can be not always random and stationary, migrating with a certain mobility governed by the activation barrier and temperature [3]. Such migration and relaxation to the equilibrium state as well as the features of the growth technology can result in a correlation or even an ordering in the configuration of point or/and line defects. Experimental observation of correlation in the spatial distribution of disorder have been reported in Ref. [7], where authors addressed enhancement of the conductivity to the effect of correlation between the potassium atoms doping the graphene. This conclusion, in turn, was based on the theoretical predictions in Ref. [8] that the correlations in the position of long-range scatterers strongly enhances the conductivity. It should be noted that the approach in Ref. [8] is based on the standard Boltzmann approach within the Born approximation. However, the applicability of the Born approximation for graphene has been questioned in Ref. [9], where it was shown that predictions based on the standard semi-classical Boltzmann approach within the Born approximation for the case of the long-range Gaussian potential are in quantitative and qualitative disagreement with the exact quantum-mechanical results in the parameter range corresponding to realistic systems. Therefore it is of interest to study the effect of spatial correlation of dopant atoms by exact numerical methods.

Correlation of extended defects, which act as line scatterers, has been also experimentally revealed [10]. Particularly, it concern the line defects in epitaxial and chemically-vapor-deposited (CVD) graphene, where they manifest correlation in their orientation and can be even almost parallel to each other due to the growth technique [10, 11, 12, 13]. Such correlation leads to the anisotropy of diffusivity and conductivity in different directions of graphene sheets [10, 13, 14].

In the present chapter, we consider different (re)distributions of point and one-dimensional ($1D$) defects in graphene lattice and then ascertain how do their configurations affect the diffusivity and hence conductivity of charge carriers. To do it, we combine analytical (statistical-thermodynamic along with kinetic) approaches and numerical (quantum-mechanical) calculations. An advantage of analytical method is account of interatomic interactions of all (but not only commonly assumed nearest-neighbor) atoms in the system. Numerical (Kubo–Greenwood-formalism-based) calculations are especially suited to treat large graphene sheets containing millions of atoms, i.e. with dimensions approaching realistic systems (here, as well as in Refs. [15, 16], we treat systems having the sizes of 1700×1000 and 3400×2000 sites corresponding to 210×210 and 420×420 nm^2). In case of the point defects, we consider random, correlated, and ordered distributions of dopant atoms, calculate the conductivity for each case, and compare obtained results. Line defects are also considered as randomly distributed and orientationally correlated, i.e. with a prevailing direction in their orientation.

Substitutional and Interstitial Graphene-Based Superstructures: Statistical-Thermodynamic Approach

Let us consider possible ordered distributions of impurity atoms over the sites and interstices of honeycomb lattice, namely, graphene-based substitutional and interstitial (super)structures, which are stable against the formation of antiphase boundaries (or splitting up onto antiphase domains).

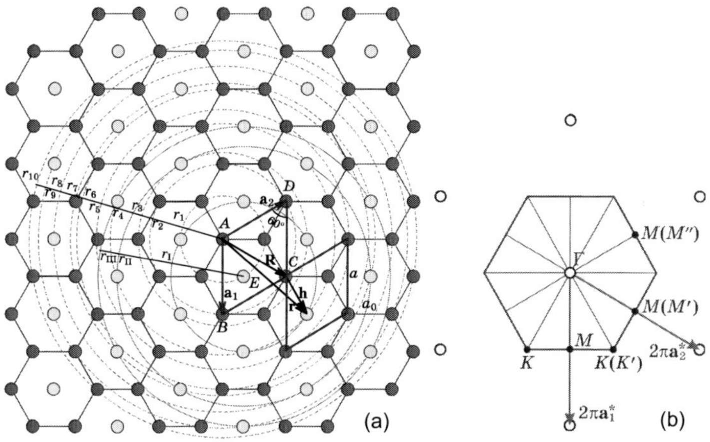

Figure 1. (Color online) (a) Crystal lattice of graphene. Here, $ABCD$ is a primitive unit cell; $\mathbf{a}_1$ and $\mathbf{a}_2$ are the basis translation vectors of the lattice; a is the lattice translation parameter; a_0 is a distance between the nearest-neighbor sites; circles (with radii r_1, r_2, ..., r_{10}, and r_I, r_{II}, r_{III}) denote the first ten substitutional (dashed line) and three interstitial (solid line) coordination shells (zones) with respect to the origin (at A site) of the oblique coordinate system and interstice E, respectively. (b) The first Brillouin zone (BZ) of the reciprocal space of graphene lattice, where Γ, M, K are its high-symmetry points; $\mathbf{a}_1^*$ and $\mathbf{a}_2^*$ are the basis translation vectors of two-dimensional reciprocal lattice.

Substitutional Superlattices

Ordered distributions of substitutional atoms A over the sites of honeycomb lattice at the stoichiometries $c_{st} = 1/2$, (CA), $1/4$ (C_3A), $1/8$ (C_7A), $1/3$ (C_2A), $1/6$ (C_5A) are shown in Fig. 2. Using the static-concentration-waves' method and the self-consistent field (mean-field) approximation [17], one can derive expressions for the configurational free energies of different honeycomb-lattice-based structures,

$$F = U - TS, \tag{1}$$

where U and S denote configurational internal energy and entropy, respectively, and T is an absolute temperature.

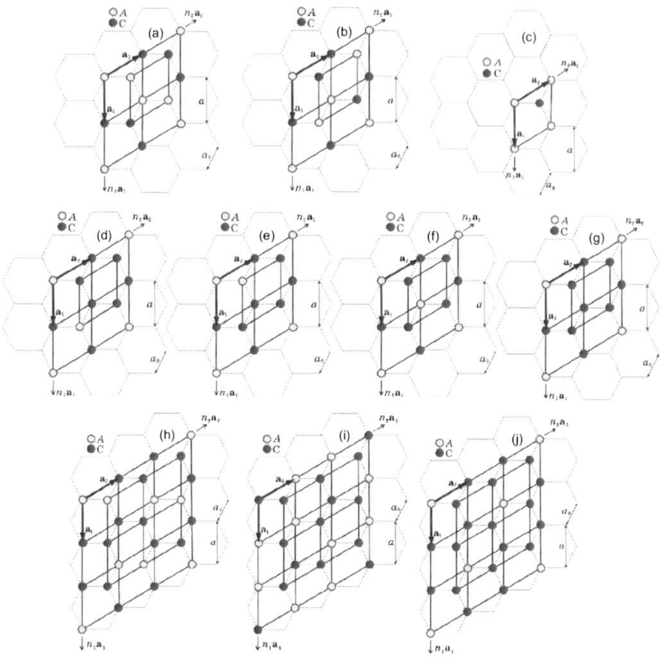

Figure 2. (Color online) Primitive unit cells of substitutional superstructures with stoichiometries $1/2$ (a)–(c), $1/4$ (d)–(f), $1/8$ (g), $1/3$ (h)–(i), $1/6$ (j).

Specific (per site) configuration-dependent part of the free energies for CA-type substitutional (super)structures in Figs. 2(a)–(c) are as follows:

$$F_1^{\mathrm{C}A} \cong \frac{1}{2}c^2\lambda_1(\mathbf{0}) + \frac{1}{8}(\eta_1^I)^2\lambda_1(\mathbf{k}^M) - TS_1^{\mathrm{C}A}(c,\eta_1^I), \quad (2)$$

$$F_2^{\mathrm{C}A} \cong \frac{1}{2}c^2\lambda_1(\mathbf{0}) + \frac{1}{8}(\eta_2^I)^2\lambda_2(\mathbf{k}^M) - TS_2^{\mathrm{C}A}(c,\eta_2^I), \quad (3)$$

$$F_3^{\mathrm{C}A} \cong \frac{1}{2}c^2\lambda_1(\mathbf{0}) + \frac{1}{8}(\eta_0^I)^2\lambda_2(\mathbf{0}) - TS_3^{\mathrm{C}A}(c,\eta_0^I). \quad (4)$$

Configurational free energies (per site) for C_2A-type substitutional (super)structures presented in Figs. 2(h) and (i) are

$$F_1^{\mathrm{C}_2A} \cong \frac{1}{2}c^2\lambda_1(\mathbf{0}) + \frac{1}{9}(\eta_1^I)^2\lambda_2(\mathbf{k}^K) - TS_1^{\mathrm{C}_2A}(c,\eta_1^I). \tag{5}$$

$$F_3^{\mathrm{C}_2A} \cong \tfrac{1}{2}c^2\lambda_1(\mathbf{0}) + \tfrac{1}{18}(\eta_0^{III})^2\lambda_2(\mathbf{0}) + \tfrac{1}{36}\left[(\eta_1^{III})^2 + (\eta_2^{III})^2\right]\lambda_2(\mathbf{k}^K) - \\ -TS_3^{\mathrm{C}_2A}(c,\eta_0^{III},\eta_1^{III},\eta_2^{III}). \tag{6}$$

Configurational free energies (per site) for C_3A-type substitutional (super)structures presented in Figs. 2(d)–(f) are

$$F_1^{\mathrm{C}_3A} \cong \frac{1}{2}c^2\lambda_1(\mathbf{0}) + \frac{3}{32}(\eta_2^I)^2\lambda_2(\mathbf{k}^M) - TS_1^{\mathrm{C}_3A}(c,\eta_2^I), \tag{7}$$

$$F_2^{\mathrm{C}_3A} \cong \frac{1}{2}c^2\lambda_1(\mathbf{0}) + \frac{1}{32}\left[2(\eta_1^{II})^2\lambda_1(\mathbf{k}^M) + (\eta_2^{II})^2\lambda_2(\mathbf{k}^M)\right] - TS_2^{\mathrm{C}_3A}(c,\eta_1^{II},\eta_2^{II}), \tag{8}$$

$$F_3^{\mathrm{C}_3A} \cong \tfrac{1}{2}c^2\lambda_1(\mathbf{0}) + \tfrac{1}{32}\left[(\eta_0^{III})^2\lambda_2(\mathbf{0}) + (\eta_1^{III})^2\lambda_1(\mathbf{k}^M) + (\eta_2^{III})^2\lambda_2(\mathbf{k}^M)\right] - \\ -TS_3^{\mathrm{C}_3A}(c,\eta_0^{III},\eta_1^{III},\eta_2^{III}). \tag{9}$$

Configurational free energy (per site) for C_5A-type substitutional (super)structure [Fig. 2(j)] reads as

$$F^{\mathrm{C}_5A} \cong \tfrac{1}{2}c^2\lambda_1(\mathbf{0}) + \tfrac{1}{72}(\eta_0^{III})^2\lambda_2(\mathbf{0}) + \tfrac{1}{36}\left[(\eta_1^{III})^2 + (\eta_2^{III})^2\right]\lambda_2(\mathbf{k}^K) - \\ -TS^{\mathrm{C}_5A}(c,\eta_0^{III},\eta_1^{III},\eta_2^{III}). \tag{10}$$

At last, configurational free energy (per site) for C_7A-type substitutional (super)structure [Fig. 2(g)]:

$$F^{\mathrm{C}_7A} \cong \tfrac{1}{2}c^2\lambda_1(\mathbf{0}) + \tfrac{1}{128}\left[(\eta_0^{III})^2\lambda_2(\mathbf{0}) + 3(\eta_1^{III})^2\lambda_1(\mathbf{k}^M) + 3(\eta_2^{III})^2\lambda_2(\mathbf{k}^M)\right] - \\ -TS^{\mathrm{C}_7A}(c,\eta_0^{III},\eta_1^{III},\eta_2^{III}). \tag{11}$$

Here, in Eqs. (2)–(11), c is an atomic fraction of dopant atoms (A), $\eta_\varsigma^\aleph$ ($\varsigma = 0$, 1 or 2) are the long-range order (LRO) parameters (Ξ index denotes their total number for a given structure; $\aleph = I, II$ or III), $\mathbf{k}$ is a wave vector belonging to the first Brillouin zone of the honeycomb-lattice reciprocal space [see Fig. 1(b)] and generating certain superstructure [in detail see Refs. [18, 19, 20, 21]]. Al the thermodynamics of the honeycomb-lattice-based superstructures in Fig. 2 is defined by interatomic-interaction parameters $\lambda_1(\mathbf{0})$, $\lambda_2(\mathbf{0})$, $\lambda_1(\mathbf{k}^M)$, $\lambda_2(\mathbf{k}^M)$, $\lambda_2(\mathbf{k}^K)$, which are connected with pairwise interaction energies $W_{pq}^{\mathrm{CC}}(\mathbf{R}-\mathbf{R}')$, $W_{pq}^{AA}(\mathbf{R}-\mathbf{R}')$, $W_{pq}^{\mathrm{C}A}(\mathbf{R}-\mathbf{R}')$ between C–C, A–A, C–A atoms situated at the sites of p-th and q-th (p, $q = 1$, 2) sublattices within the unit sells with origins ("zero" sites) at $\mathbf{R}$ and $\mathbf{R}'$. Pairwise interaction energies define so-called mixing energy, $w_{pq}(\mathbf{R}-\mathbf{R}') \equiv W_{pq}^{\mathrm{CC}}(\mathbf{R}-\mathbf{R}') + W_{pq}^{AA}(\mathbf{R}-\mathbf{R}') - 2W_{pq}^{\mathrm{C}A}(\mathbf{R}-\mathbf{R}')$, which in the literature is known also as "ordering energy" and "interchange energy" [17, 20, 21]. The first-, second-, ..., n-th-neighbor mixing energies, w_1, w_2, ..., w_n [see Fig. 1(a)], are commonly used for the analysis of the equilibrium atomic order [19, 21, 22] as well as the ordering kinetics [18, 20, 23]. For the statistical-thermodynamic description of the interatomic interactions in all coordination shells, or arbitrary-range interactions, it is conveniently to apply the Fourier transform, which results to the interatomic-interaction parameters $\lambda(\mathbf{k})$ entering into the expressions for the configuration free energies (2)–(11).

Interstitial Superlattices

Denote interstitial atoms in graphene lattice as X, and remained vacant positions for these atoms in the interstices as $\varnothing$. Each primitive unit cell of the honeycomb lattice contains two sites and one interstice being center of the comb [Fig. 1(a)]. An occupation of all interstices by the dopant atoms X corresponds to the relative impurity concentration $\kappa = \kappa_{st} = 1$ and results to the superstructure-cluster C_2X with a maximal atomic fraction of the interstitial dopant atoms, $c = c_{st} = 1/3$. Its primitive unit cell is shown in Fig. 3(a). In this case, applying the static concentration waves' approach and the self-consistent field (mean-field) approximation [17], one get distribution function for impurity atoms, $P(\mathbf{R}) \equiv 1$, and specific configurational free energy (i.e. energy per one interstice) $F^{\mathrm{C}_2X} \cong w(\mathbf{0})/2$. Here $w(\mathbf{k}=\mathbf{0})$ is a Fourier-transform of the mixing energy of X and $\varnothing$ components of interstitial subsystem, $w(\mathbf{R}-\mathbf{R}') \equiv W^{XX}(\mathbf{R}-\mathbf{R}') + W^{\varnothing\varnothing}(\mathbf{R}-\mathbf{R}') - 2W^{X\varnothing}(\mathbf{R}-\mathbf{R}')$, where $W^{\alpha\beta}(\mathbf{R}-\mathbf{R}')$ is energy of effectively pairwise interaction of α and β ($\alpha, \beta = X, \varnothing$) kinds of "atoms" occupying interstices within the primitive unit cells with radii-vectors $\mathbf{R}$ and $\mathbf{R}'$, respectively.

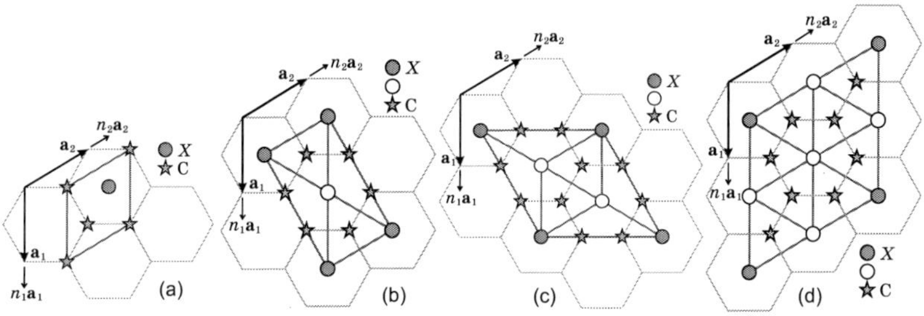

Figure 3. (Color online) Primitive unit cells of interstitial superstructures with stoichiometries $1/3$ (a), $1/5$ (b), $1/7$ (c), $1/9$ (d). Stars denote carbon atoms, and open circles—unoccupied (in the superstructure) interstices.

Specific (per interstice) configurational free energy of C_4X-type interstitial (super)structure [Fig. 3(b)], where in totally ordered state (at 0 K) the relative concentration of interstitial atoms $\kappa_{st} = 1/2$, i.e. their atomic fraction $c_{st} = 1/5$, reads as

$$F^{C_4X} \cong \frac{1}{2}\kappa^2\widetilde{w}(\mathbf{0}) + \frac{1}{8}\eta^2\widetilde{w}(\mathbf{k}^M) - TS^{C_4X}(\kappa,\eta). \tag{12}$$

Configurational free energy (per interstice) of C_6X-type interstitial (super)structure in Fig. 3(c) (in the totally ordered state $\kappa_{st} = 1/3$, $c_{st} = 1/7$) is

$$F^{C_6X} \cong \frac{1}{2}\kappa^2\widetilde{w}(\mathbf{0}) + \frac{1}{9}\eta^2\widetilde{w}(\mathbf{k}^K) - TS^{C_6X}(\kappa,\eta). \tag{13}$$

In the totally ordered state of C_8X-type interstitial (super)structure [Fig. 3(d)], $\kappa_{st} = 1/4$ and $c_{st} = 1/9$. Its specific configuration-dependent part of the free energy reads as

$$F^{C_8X} \cong \frac{1}{2}\kappa^2\widetilde{w}(\mathbf{0}) + \frac{3}{32}\eta^2\widetilde{w}(\mathbf{k}^M) - TS^{C_8X}(\kappa,\eta). \tag{14}$$

In conclusion of this section, note that all free-energy equations (2)–(14), being derived within the framework of the self-consistent field approximation, are "governed" by the effective pairwise interactions of α–β particles only, where $\alpha, \beta = \mathrm{C}, A$ for substitutional systems, and $\alpha, \beta = X, \varnothing$ for interstitial ones. The main point of such model is that total internal field acting on the substitutional (interstitial) "atom" from the other substitutional (interstitial) and matrix atoms, is replaced with self-averaged (self-consistent) field representing the most probable result of the total interaction of all atoms with distinguished one for a given their distribution generated by the same (self-consistent) field [17, 25, 26].

Low-Temperature Stability of Superstructures

Since each of the stoichiometries among the interstitial superstructures predicts only one ordered distribution of interstitial atoms, as Figs. 3(a)–(d) demonstrate, below we pay an attention to the substitutional (super)structural stability only. (Peculiarities of the stability of interstitial graphene-based structures can be found in Ref. [26].)

As follows from Eq. (1), the low-temperature (i.e. at $T = 0$ K) stability of a structure, when contribution of the entropy, S, to the total free energy, F, is small, depends on the internal energy, U. At $T = 0$ K, the stable is a phase which has the lowest internal energy as compared with other phases of the same composition (here we are neglecting the possibility of the formation of mechanical mixture of pure components and different structures). So, to calculate the low-temperature stability ranges, we minimize the configurational free energy, $F = U|_{T=0}$, setting in Eqs. (2)–(11) $T = 0$. Such minimization is a sufficient stability condition. The necessary condition any superstructure to be appeared is a positive temperature of the stability loss of disordered state with respect to appearance of the long-range atomic order: $T_s = -k_B^{-1}c(1-c)\lambda_\omega(\mathbf{k}) > 0$, i.e., first of all, $\min\lambda_\omega(\mathbf{k}) < 0$ ($\omega = 1, 2$; $k \in BZ$) [17]. These two (necessary and sufficient) conditions can be realized in a certain range of interatomic-interaction parameters $\lambda(\mathbf{k})$ entering into Eqs. (2)–(11).

The CA-, C_2A-, and C_3A-type superstructures seem the most interesting, since, at these stoichiometries, there are three or two different (nonequivalent) ordered distributions of atoms (see Fig. 2). The low-temperature stability regions for these superstructures are represented in Figs. 4 and 5, where the ranges of values of interatomic-interaction parameters providing such a stability are determined. Two cases are considered: firstly (Fig. 4), taking into account only first-, second- and third-neighbor mixing energies (w_1, w_2, w_3), but vanishing mixing energies in other (distant) coordination shells, and, secondly (Fig. 5),

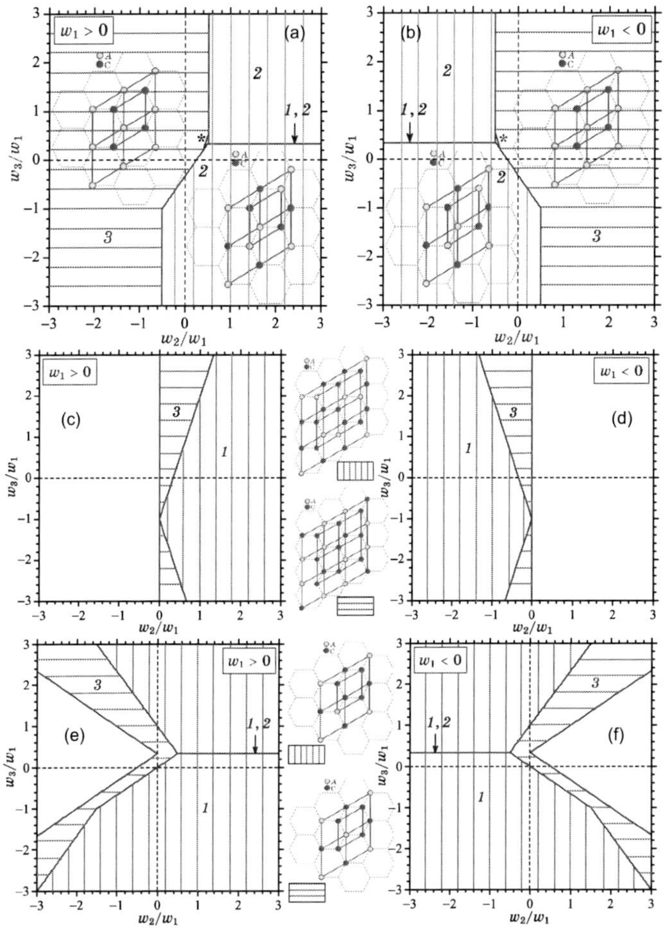

Figure 4. (Color online) The low-temperature stability regions (in terms of the ratios of the mixing energies w_2/w_1 and w_3/w_1) for CA (a), (b); C_2A (c), (d); C_3A (e), (f) superstructures assuming interatomic interactions in the first three coordination shells. Here, (a)–(b) *1*, *2*, *3* denote $\lambda_1(\mathbf{k}^M)$, $\lambda_2(\mathbf{k}^M)$, $\lambda_2(\mathbf{0})$ entering into Eqs. (2)–(4) for CA; (c)–(f) *1*, *2*, *3* denote number of LRO parameters describing C_2A and C_3A.

taking into account mixing energies in all coordination shells.

An account of the third-nearest-neighbor interatomic interactions always provides the stability for the superstructures [Figs. 2(c), (d), (f)] in which substitutional dopant atoms are surrounded by the opposite-kind neighbors. However, an account of (only) these (short-range) interactions can be an inadequate to provide the stability for the superstructures [Figs. 2(a) and (e)] in which some of the dopant atoms occupy the nearest-neighbor lattice sites. Figure 5 demonstrates that accounting of the interactions of all atoms contained in the system yields new results as compared with those obtained within the scope of the third-nearest-neighbor interaction approach: every predicted superstructures can be stable at the appropriate values of interatomic-interaction energies.

At the stoichiometries 1/8 and 1/6, there is only one possible ordered arrangement of atoms [see Figs. 2(g), (j) and also Eqs. (10), (11)]. Therefore, at low temperatures, C_7A- and C_5A-type honeycomb-lattice-based superstructures are stable in all set of interatomic-interaction-energy values.

Thus, the third-nearest-neighbor Ising model results in the instability (thermodynamic unfavorableness) of some predicted superstructures. In contrast to this model, the consideration of all coordination shells in the interatomic interactions shows that all predicted honeycomb-lattice-based superstructures are stable at the appropriate values of interatomic-interaction energies. Moreover, some superstructures [CA and C_3A in Figs. 2(a) and (e), respectively] practically may be stable due to the long-range interatomic interactions only.

The problem of stability for graphene-based structures is considered at low temperatures. At finite (or room) temperatures, when LRO parameters in Eqs. (2)–(11) are not equal to unity, $\eta_{\varsigma}^{\aleph} \neq 1$, an entropy contribution to the free energy appears. It will result in a shift of the boundaries between the stability ranges in Figs. 4 and 5, but it will not change the qualitative results, particularly, the long-range interatomic-interaction effect on the stability of the graphene-based (super)structures.

Kinetics of the Long-Range Atomic-Order Relaxation

As it is shown above, all interstitial (super)structures (Fig. 2) are described by the one LRO parameter only [Eqs. (12)–(14)], while this is not the case for substitutional ones (Fig. 3), where two and even there LRO parameters can enter

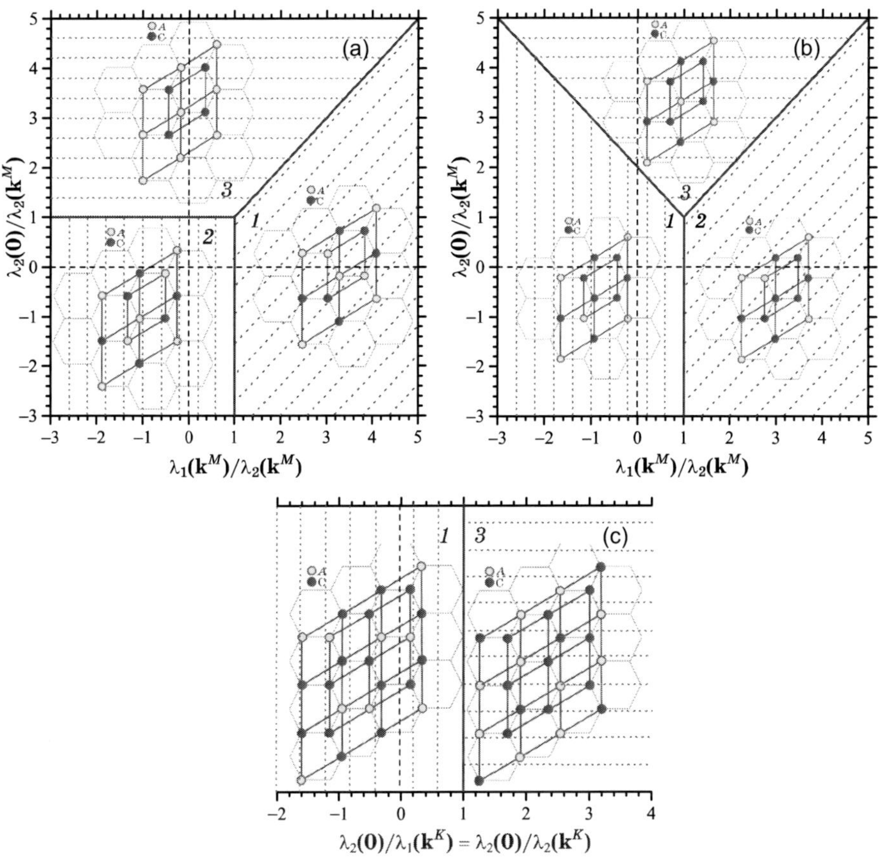

Figure 5. (Color online) The same as in the previous figure, but taking into account interactions of all atoms in the superstructures CA (a), C_3A (b), and C_2A (c).

into the free-energy equations (2)–(11). That is why here we consider more complex case—kinetics of the LRO relaxation in the substitutional systems. (Details of the LRO relaxation in the interstitial graphene-based systems can be found in Ref. [26].)

Lets us describe the long-range atomic-order kinetics considering case of exchange (ring) diffusion mechanism governing the atomic ordering in a two-dimensional binary solid solution $\mathrm{C}_{1-c}A_c$ based on a graphene-type lattice (neglecting the vacancies at the lattice sites). Apply the Önsager-type microdiffusion master equation [17, 20, 23]:

$$\frac{dP_p^\alpha(\mathbf{R},t)}{dt} \approx -\frac{1}{k_BT}\sum_{\beta=\mathrm{C},A}\sum_{q=1}^{2}\sum_{\mathbf{R}'} c_\alpha c_\beta L_{pq}^{\alpha\beta}(\mathbf{R}-\mathbf{R}')\frac{\delta\triangle F}{\delta P_q^\beta(\mathbf{R}',t)}; \quad (15)$$

here, $P_p^\alpha(\mathbf{R},t)$ is a probability to find α-atom in a time t at the $(p,\mathbf{R})$ site, i.e. at the site of q-th sublattice within the unit-cell origin position $\mathbf{R}$; c_α (c_β) is a relative fraction of α-kind (β-kind) atom; $||L_{pq}^{\alpha\beta}(\mathbf{R}-\mathbf{R}')||$ is a matrix of the Önsager-type kinetic coefficients whose elements represent probabilities of elementary exchange-diffusion jumps of a pair of α and β atoms at $\mathbf{r}=\mathbf{R}+\mathbf{h}_p$ and $\mathbf{r}'=\mathbf{R}'+\mathbf{h}_q$ sites of the p-th and q-th sublattices composing the honeycomb lattice and displaced with respect to each other by the vector $\mathbf{h}$ ($\alpha,\beta=\mathrm{C},A$; $p,q=1,2$; $c_A=c$, $c_\mathrm{C}=1-c$).

If the vacancy content is small, we have almost identity for the single-site occupation-probability functions of A and C atoms distribution over the honeycomb-lattice site: $P_q^\mathrm{C}(\mathbf{R},t)+P_q^A(\mathbf{R},t)\approx 1\ \forall\mathbf{R}\wedge\forall q=1,2\wedge\forall t>0$. Then it is enough to consider an exchange-microdiffusion migration of only dopant atoms A in terms of the time dependence of only probabilities $\{P_q(\mathbf{R},t)\}$ $[P_q(\mathbf{R},t)\equiv P_q^A(\mathbf{R},t)\ \forall t>0]$. One can use kinetics equation (15) to describe microdiffusion by the other mechanisms since semi-phenomenological Eq. (15) does not contain a certain microdiffusion mechanism. Considering of any other mechanism does not require changing of the type of Eq. (15), since diffusion mechanism is defined by the kinetic coefficients $L_{pq}^{\alpha\beta}(\mathbf{R}-\mathbf{R}')$, which should be linked with microscopic characteristics of the system (energy barrier heights for atomic jumps, thermal vibrational frequencies of atoms at the sites, vacancy concentration) and external thermodynamic parameters (temperature etc.)

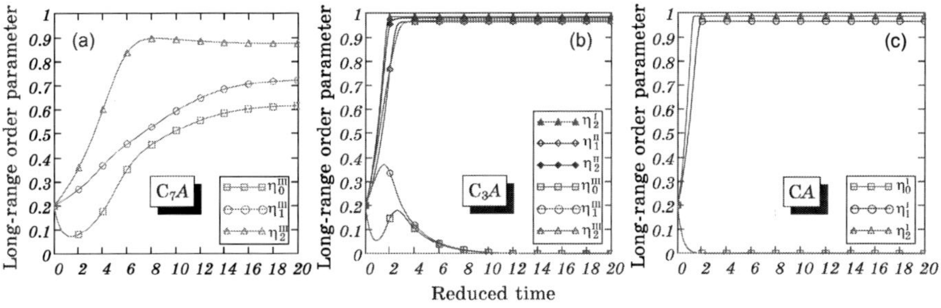

Figure 6. (Color online) The time evolution of the LRO parameters in the graphene-based systems for the temperature $T^* = k_BT/|\lambda_2(\mathbf{k}^M)|$ and interatomic-interaction parameters parameters $\lambda_1(\mathbf{k}^M)/\lambda_2(\mathbf{k}^M) = 5/6$, $\lambda_2(\mathbf{0})/\lambda_2(\mathbf{k}^M) = -5/8$ ($\lambda_2(\mathbf{k}^M) < 0$).

Condition of the conservation of each kind of atoms in the system, assumption that any site is obligatory occupied by C or A atom, Önsager-type symmetry relations, representation of thermodynamic driven force $\delta\triangle F/\delta P_q(\mathbf{R}')$ (as well as $P_p(\mathbf{R})$) as a superposition of the concentration waves, followed by the Fourier transform of both members in Eq. (15), yield us differential equations of the time evolution of the LRO parameters, $\eta_\varsigma^\aleph$:

$$\frac{d\eta_\varsigma^\aleph}{dt} \cong -c(1-c)\widetilde{L}(\mathbf{k})\left[\eta_\varsigma^\aleph \frac{\lambda_\omega(\mathbf{k})}{k_BT} + \ln Z(c, \eta_0^\aleph, \eta_1^\aleph, \eta_2^\aleph)\right], \qquad (16)$$

where $\widetilde{L}(\mathbf{k})$ is the Fourier-component of a concentration-dependent combination of kinetic coefficients $L_{pq}^{\alpha\beta}(\mathbf{R}-\mathbf{R}')$, $\widetilde{L}_{pq}^{\alpha\beta}(\mathbf{k}) \equiv \sum_{\mathbf{R}} L_{pq}^{\alpha\beta}(\mathbf{R}-\mathbf{R}')\exp[-i\mathbf{k}\cdot(\mathbf{R}-\mathbf{R}')]$, and particular expressions for $Z(c, \eta_0^\aleph, \eta_1^\aleph, \eta_2^\aleph)$ are presented in Refs. [20, 23]. It is convenient to solve Eq. (16) in terms of the reduced time $t^* = \widetilde{L}(\mathbf{k})t$ and temperature $T^* = k_BT/|\lambda_\omega(\mathbf{k})|$.

Curves in Fig. 6 are numerical calculations of the kinetic equations (16) for the ordered C_7A, C_3A, and CA superstructural types at the reduced temperature $T^* = 0.1$ and certain interatomic-interaction parameters $\lambda_\omega(\mathbf{k})$, given as an example. These values correspond to the certain point [(5/6, −5/8)] on the stability diagrams for CA and C_3A superstructures in Figs. 5(a) and (b). This point indicates what superstructure is energetically favorable (stable) between the three possible ones at the given stoichiometry. Stability diagrams in Fig. 5

are obtained for the absolute zero temperature, while the kinetic curves in Fig. 6 are calculated for the nonzero temperature. Nevertheless, one can easy see a correspondence between the statistical-thermodynamic and kinetics results. Results of the latter improve previous ones; particularly, for the mentioned point on the diagrams, energetically favorable is a structure described by the LRO parameter, which relaxes to its equilibrium state being the highest between the other equilibrium, stationary, and current values of the LRO parameters of the given composition (see Figs. 5(a), (b) and Figs. 6(b), (c).

Figures 6(a) and (b) clear demonstrate that kinetic curves for the LRO parameters of the C_7A- and C_3A-type (super)structures, described by two or three order parameters, can be nonmonotonic. The nonmonotony is caused not only by the presence of two interpenetrating sublattices composing the honeycomb lattice, but also by the dominance of the intersublattice mixing (interatomic-interaction) energies in their competition with intrasublattice interaction energies.

Influence of Correlated and/or Ordered Impurities on Conductivity of Graphene: Numerical Calculations

This section is devoted to the investigation of influence of the spatial correlation and ordering of impurities, acting as a "disorder" in graphene, on its conductance using a numerical quantum mechanical approach. We utilize the time-dependent real-space quantum Kubo–Greenwood method [15, 16, 27, 28, 29, 30, 31, 32, 33], which allows us to study experimentally-relevant large graphene sheets containing millions of atoms. We consider models of disorder appropriate for realistic impurities that might exhibit correlations, including the Gaussian potential describing screened charged impurities and the short-range potential describing neutral adatoms.

We model electron dynamics in graphene using the standard p-orbital nearest neighbor tight-binding Hamiltonian defined on a honeycomb lattice [5, 6],

$$\hat{H} = -u \sum_{i,i'} c_i^\dagger c_{i'} + \sum_i V_i c_i^\dagger c_i, \tag{17}$$

where $c_i^\dagger$ and c_i are the standard creation and annihilation operators acting on a quasi-particle on the site i. The summation over i runs over the entire graphene lattice, while i' is restricted to the sites next to i; $u = 2.7$ eV is the hopping integral for the neighboring C atoms i and i' with distance $a_0 \approx 0.142$ nm between them (Fig. 1), and V_i is the on-site potential describing impurity scattering.

In the present study we consider both short- and long-range impurities. The short-range impurities represent neutral adatoms on graphene and are modeled by the delta-function scattering potential for electrons

$$V_i = \sum_{j=1}^{N_{imp}} V_j \delta_{ij}, \tag{18}$$

where N_{imp} is the number of impurities on a graphene sheet. Estimations based on *ab initio* calculations and the T-matrix approach for common adatoms provide typical values for the on-site potential $V_j = V_0 \lesssim 80u$ [34, 35, 36, 37], e.g., for hydrogen, $V_0 \approx 60u$.

The long-range potential is appropriate for screened charged impurities situated on graphene and/or dielectric substrate. We model them by the Gaussian scattering potential commonly used in the literature [5, 6, 9]

$$V_i = \sum_{j=1}^{N_{imp}} U_j \exp\left(-\frac{|\mathbf{R}_i - \mathbf{R}_j|^2}{2\xi^2}\right), \tag{19}$$

where $\mathbf{R}_i$ ($\mathbf{R}_j$) is the radius-vector of the i (j) site, ξ is the effective potential radius, and the potential height is uniformly distributed in the range $U_j \in [-\Delta, \Delta]$ with Δ being the maximum potential height.

We consider three cases of impurity distribution, random (uncorrelated), correlated, and ordered. In the first case, the summation in Eqs. (18), (19) is performed over the random distribution of impurities over the lattice sites. In the second case impurities are no longer considered to be randomly distributed and to describe their spatial correlation we adopt a model used in Ref. [8] introducing the pair distribution function $P(\mathbf{R}_i - \mathbf{R}_j) \equiv P(r)$,

$$P(r) \equiv \begin{cases} 0, & r < r_0 \\ 1, & r > r_0 \end{cases} \tag{20}$$

where $r = |\mathbf{R}_i - \mathbf{R}_j|$ is the distance between two impurities and the correlation length r_0 defines the minimum distance that can separate two impurities. Note, that for the randomly distributed (totally uncorrelated) impurities

$r_0 = 0$. The largest distance $r_{0_{\max}}$ depends on the relative impurity concentration c: the smaller the concentration c, the larger $r_{0_{\max}}$. At last, in the third case, the ordered distribution of impurity atoms over the sites of both sublattices is described by the single-site occupation-probability functions, which can be derived by the method of the static concentration waves [17]. Particularly, for C_7A-type ($c_{st} = 1/8$) substitutional superstructure in Fig. 2(g), they are [18, 19, 20]

$$\begin{pmatrix} P_1(\mathbf{R}) \\ P_2(\mathbf{R}) \end{pmatrix} = c\begin{pmatrix} 1 \\ 1 \end{pmatrix} + \frac{1}{8}\eta_0^{III}\begin{pmatrix} 1 \\ -1 \end{pmatrix} +$$
$$+\frac{1}{8}\eta_1^{III}\left[\begin{pmatrix} 1 \\ 1 \end{pmatrix}\cos\pi n_1 + \begin{pmatrix} 1 \\ -1 \end{pmatrix}\cos\pi n_2 + \begin{pmatrix} 1 \\ 1 \end{pmatrix}\cos\pi(-n_1+n_2)\right] +$$
$$\frac{1}{8}\eta_2^{III}\left[\begin{pmatrix} 1 \\ -1 \end{pmatrix}\cos\pi n_1 + \begin{pmatrix} 1 \\ 1 \end{pmatrix}\cos\pi n_2 + \begin{pmatrix} 1 \\ -1 \end{pmatrix}\cos\pi(-n_1+n_2)\right]; \tag{21}$$

n_1, n_2, n_3 are integers. $P_1(\mathbf{R})$ and $P_2(\mathbf{R})$ possess four values $[c + \frac{1}{8}(\eta_0^{III} + 3\eta_1^{III} + 3\eta_2^{III})$, $c + \frac{1}{8}(\eta_0^{III} - \eta_1^{III} - \eta_2^{III})$, $c + \frac{1}{8}(-\eta_0^{III} + \eta_1^{III} - \eta_2^{III})$, $c + \frac{1}{8}(-\eta_0^{III} - 3\eta_1^{III} + 3\eta_2^{III})]$ over all lattice sites. The representative examples of random and correlated distributions for the cases of the short- and long-range potentials are shown in Fig. 7.

The transport properties of graphene sheets can be calculated on the basis of the time-dependent real-space Kubo formalism, extracting the dc conductivity σ from the wave packet temporal dynamics governed by the time-dependent Schrödinger equation [15, 16, 27, 28, 29, 30, 31, 32, 33].

A central quantity in the Kubo–Greenwood approach is the mean quadratic spreading of the wave packet along the x-direction at the energy E, $\Delta\hat{X}^2(E,t) = \langle(\hat{X}(t) - \hat{X}(0))^2\rangle$, where $\hat{X}(t) = \hat{U}^\dagger(t)\hat{X}\hat{U}(t)$ is the position operator in the Heisenberg representation, and $\hat{U}(t) = e^{-i\hat{H}t/\hbar}$ is the time-evolution operator [wave-packet propagation is visualized in Figs. 8(a), (b)]. Starting from the Kubo–Greenwood formula for the dc conductivity [38]

$$\sigma = \frac{2\pi\hbar e^2}{\Omega}\mathrm{Tr}[\hat{v}_x\delta(E-\hat{H})\hat{v}_x\delta(E-\hat{H})], \tag{22}$$

where $\hat{v}_x$ is the x-component of the velocity operator, E is the Fermi energy, Ω is the area of the graphene sheet, and factor 2 accounts for the spin degeneracy,

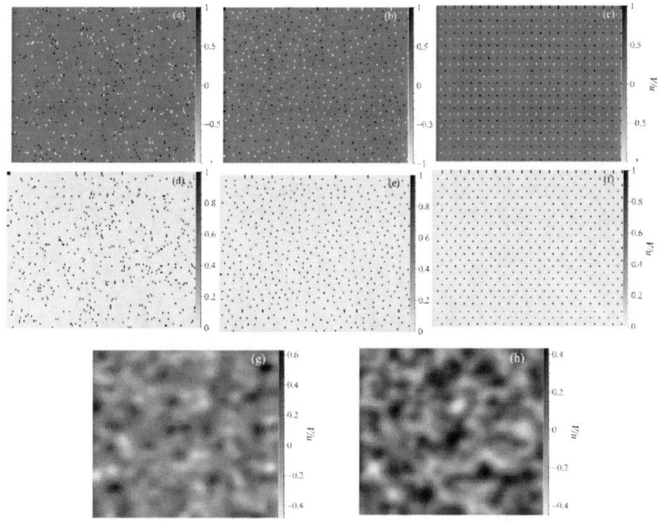

Figure 7. (Color online) A representative illustration of (a), (d), (g) random, correlated (b), (e), and (c), (f), (h) ordered distributions of impurities for short-range (a)–(c) symmetric (attractive–repulsive) and (d)–(f) asymmetric (repulsive) scattering potentials, and (g), (d) long-range Gaussian (attractive–repulsive) potential.

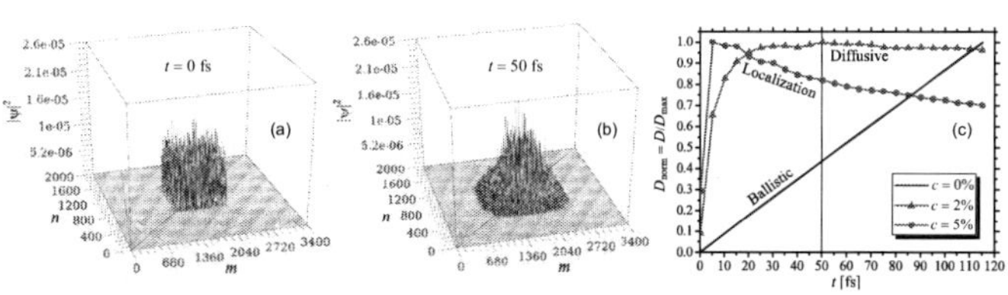

Figure 8. (Color online) Wave-packet propagation (a) and (b) in graphene lattice without ($c = 0\%$) and with short-range strong impurities ($c = 2\%$ and 5%) modeled by the onsite potential $V \sim 37u$; temporal evolution of the $D_{\mathrm{norm}}(E,t) = D(E,t)/D_{\max}(E)$ value normalized by the diffusion coefficient $D_{\max}(E)$ (c). Transport curves are presented for $E = 0.2u$.

the conductivity can then be expressed as the Einstein relation,

$$\sigma \equiv \sigma_{xx} = e^2 \tilde{\rho}(E) \lim_{t \to \infty} D(E, t), \tag{23}$$

where $\tilde{\rho}(E) = \frac{\rho}{\Omega} = \frac{\mathrm{Tr}[\delta(E-\hat{H})]}{\Omega}$ is the density of electronic sates (DOS) per unit area (per spin), and the time-dependent transport coefficient $D(E,t)$ (commonly called as diffusivity) relates to $\Delta\hat{X}^2(E,t)$ as

$$D(E,t) = \frac{\langle \Delta \hat{X}^2(E,t) \rangle}{t} = \frac{1}{t} \frac{\mathrm{Tr}[(\hat{X}_H(t) - \hat{X}(0))^2 \delta(E - \hat{H})]}{\mathrm{Tr}[\delta(E - \hat{H})]}. \tag{24}$$

Further, we are interested in the diffusive transport regime at $t \to \infty$ [Fig. 8(c)], when, neglecting the quantum-localization effects, the coefficient $D(E,t)$ reaches its maximum. Therefore, following Refs. [31, 32], we replace in Eq. (23) $\lim_{t\to\infty} D(E,t) \to D_{\max}(E)$, such that the diffusion-controlled dc conductivity is defined as

$$\sigma = e^2 \tilde{\rho}(E) D_{\max}(E). \tag{25}$$

Note that in most experiments, the conductivity is measured as a function of electron density n_e. We calculate the electron density as $n_e(E) \equiv n_e = {}^{E}_{-\infty}\tilde{\rho}(E)dE - n_{\text{ions}}$, where $n_{\text{ions}} = 3.9 \cdot 10^{15}$ cm^{-2} is the density of the positive ions in the graphene lattice compensating the negative charge of the p-electrons [for the ideal graphene lattice, i.e. without defects, at the neutrality point $n_e(E) = 0$]. Combining the calculated $n_e(E)$ with $\sigma(E)$ given by Eq. (25), one can obtain the required dependence of the conductivity $\sigma = \sigma(n_e)$. Details of numerical calculations of DOS, $D(E,t)$, and σ are given in Ref. [15].

Figure 9 shows the electron-density dependencies of the conductivity $\sigma = \sigma(n_e)$ for random and correlated impurities modeled by different scattering potentials, where positive and negative values of n_e correspond to different kinds of charge carriers: electrons and holes. As Figure 9 demonstrates,t for the most important, experimentally relevant cases of point defects, namely the strong short-range potential and the long-range Gaussian potential, the correlation in the distribution of impurity atoms does not affect the conductivity of the graphene as compared to the case when they are distributed randomly. This represent the main result for the case of the correlation. We find that the correlations lead to the enhancement of the conductivity only for the case of the weak

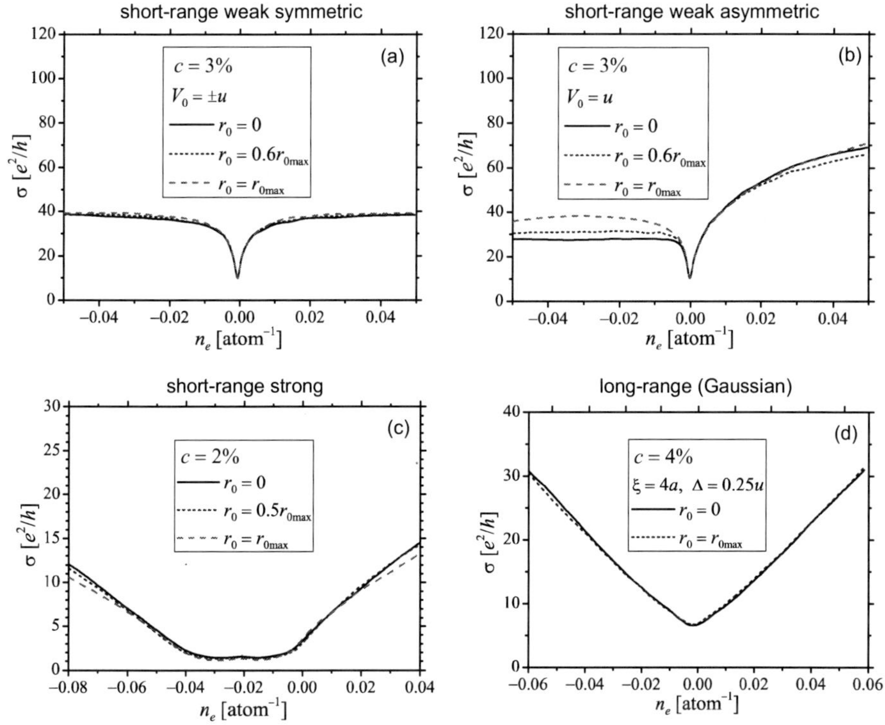

Figure 9. (Color online) Conductivity σ as a function of the relative charge carrier (electron) density n_e (the number of electrons per C atoms) for different concentration c of random ($r_0 = 0$) and correlated ($r_0 = \frac{1}{2}r_{0_{\max}}$, $r_0 = r_{0_{\max}}$) short-range weak (a) and(b), short-range strong (c), and long-range Gaussian (d) impurities.

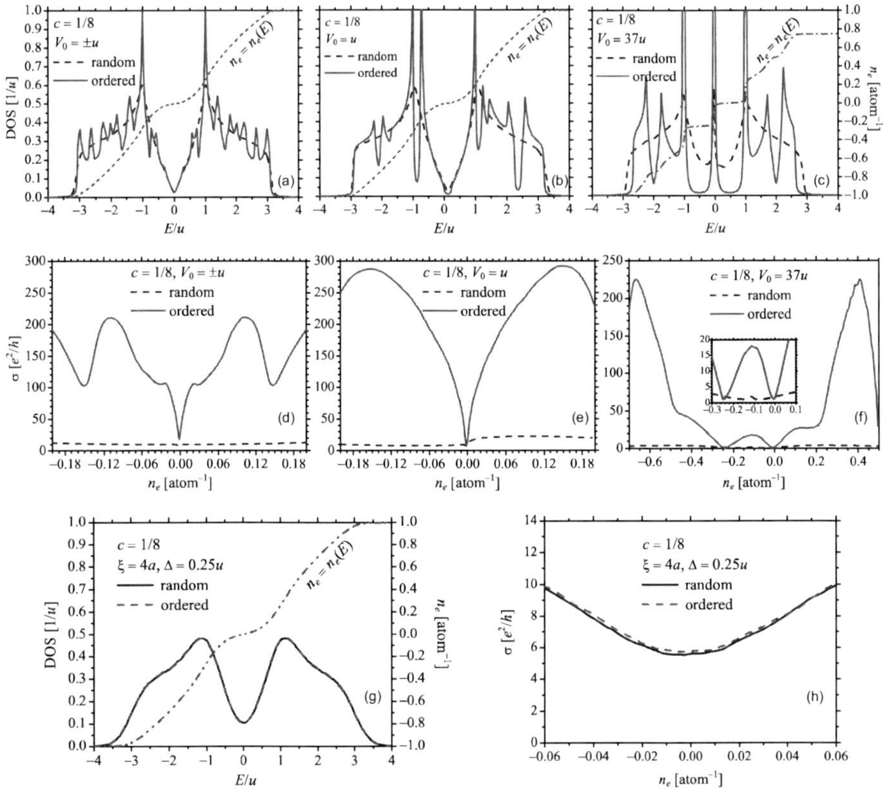

Figure 10. (Color online) Density of electronic states (a)–(c), (g) and conductivity (d)–(f), (h) for 12.5% of random and ordered impurities, modeled by the short-range weak symmetric (a) and (d), short-range weak asymmetric (b) and (e), short-range strong (c) and (f), and long-range Gaussian (g) and (h) scattering potentials.

short-range potential and only when the potential is asymmetric, i.e. $V = V_0$ or $V = -V_0$. No enhancement of the conductivity is found for the symmetric weak short-range potential, $V = \pm V_0$.

As it was mentioned in the introduction, in the recent experiment [7] the temperature increase of the conductivity was attributed to the enhancement in the spatial correlation of the adsorbed potassium ions. Numerical findings do not sustain this interpretation, the obtained here results strongly suggest that the enhancement of the conductivity reported in Ref. [7] is most likely caused by other factors not related to the correlations of impurities. The numerical calculations do not support also theoretical predictions in Ref. [8] that the correlations in the impurity positions for the long-range potential lead to the enhancement of the conductivity. This can be attributed to the utilization of the standard Boltzmann approach within the Born approximation which is not valid for the case of a long-range potential in the parameter range corresponding to realistic systems.

In contrast to the case of the correlated impurities, the ordered short-range weak and essentially strong impurities can strongly affect charge transport in graphene [Fig. 10(a)–(f)]. In the DOS-curves discrete energy levels appear [Figs. 10(a)–(c)] and broaden as impurity concentration or/and scattering potential increases. This oscillations (peaks) in the DOS and therefore in the conductivity are caused by the strongly periodic potential, describing periodic positions of impurity atoms "precipitating" a certain superstructure (in the given case, with the stoichiometry $c_{st} = 1/8$). It is clear seen from Fig. 10 that the conductivity curves for the ordering case rise up to tens times for the short-range weak [Figs. 10(d), (e)] and especially strong [Fig. 10(f)] scatterers as compared with the case of the randomly-distributed scattering centers. However, as for the correlation, ordering does not affect nor density of electronic states nor conductivity for the long-range (Gaussian) scattering potential (Figs. 10(g), (h).

Anisotropy and Increase in Conductivity due to the Orientational Correlation of Line Defects in Graphene

Nowadays, several techniques are capable of producing high-quality, large-scale graphene. These include CVD-grown graphene on transition metal surfaces [39] and epitaxial graphene growth on SiC [40]. Usually, the growth of graphene by the CVD-method requires to use metal surfaces with hexagonal symmetry, such as the (111) surface of cubic or the (0001) surface of hexagonal crystals [3]. The mismatch between the metal-substrate and graphene causes the strains in the latter, reconstructs the chemical bonds between the carbon atoms and results in formation of two-dimensional (2*D*) domains of different crystal orientations separated by one-dimensional defects [3, 41, 42, 43]. The nucleation of the graphene phase takes place simultaneously at different places, which leads to the formation of independent 2*D* domains matching corresponding grains in the substrate. A line defect appears when two graphene grains with different orientations coalesce; the stronger the interaction between graphene and the substrate, the more energetically preferable the formation of line defects is. These line defects accommodate localized states trapping the electrons, originating lines of immobile charges that scatter the Dirac fermions in graphene. It is well established that the presence of grains and grain boundaries in three-dimensional polycrystalline materials can strongly affect their electronic and transport properties. Hence, in principle, the role of such structures in 2*D* materials, such as graphene, can be even more important because even a single line defect can divide and disrupt the crystal [3]. Theoretical results [16, 44] improve that the presence of charged line defects strongly affect the transport properties of CVD-graphene. Such effect becomes more weighty due to existence of ordered line defects in CVD-synthesized graphene [13].

In epitaxial graphene the surface steps caused by substrate morphology are spatially correlated and act as line scatterers for the charge carriers [10]. Epitaxial graphene films grown on SiC [10, 12] (by SiC decomposition) or on Ru [11] (by CVD method) comprise two distinct self-organized periodic regions of terrace and step, leading to ordered graphene domains [11]. Experimental measurements show an increase of the resistance with the step density [45], the step heights [46], the step bunching [47]. Also, an anisotropy of the conductivity in the parallel and perpendicular directions to the steps is revealed, which is due to to higher defect abundance in the step regions [10, 14]. Substrate steps alone increase the resistivity in several times relative to a perfect terrace [46] with the ratio of the estimated electron mobilities in the terrace and step regions being about 10:1 [10]. Despite the strong curvature of graphene in the vicinity of steps, a structural deformation contributes only little to electron scattering

[48]. For the SiC substrate, the dominant scattering mechanism is provided by the sharp potential variations in the vicinity of the step due to the electrostatic doping from the substrate strongly coupled with graphene in the step regions [48].

Several theoretical studies have been recently reported addressing transport properties of graphene with a single graphene boundary [49, 50, 51] or polycrystalline graphene with many domain boundaries [27]. On the other hand, much less attention has been paid to the effect of charge accumulation at these boundaries due to self-doping. Transport properties of graphene with $1D$ charged defects has been studied in Ref. [44] using the Boltzmann approach within the first Born approximation. It has been demonstrated that such approximation is not always applicable for the description of electron transport in graphene even at finite (non-zero) electronic densities [37, 52, 9]. Following Ref. [16], below we present results of investigations of the impact of extended charged defects in the transport properties of graphene by an exact numerical approach based on the time-dependent real-space quantum Kubo method [15, 16, 27, 28, 29, 30, 31, 32, 33], which is especially suited for experimentally-relevant systems containing millions of atoms.

Since line defects can be thought as lines of reconstructed point defects [41, 3, 42, 43], we model a $1D$ defect as point defects oriented along a fixed direction (corresponding to the line direction) in the honeycomb lattice. The electronic effective potential for a charged line within the Thomas–Fermi approximation was first obtained in Ref. [44]. If there are N_{lines} such charged lines in a graphene lattice, the effective scattering potential reads as

$$V_i = \sum_{j=1}^{N_{\text{lines}}} U_j \left[-\cos(q_{\text{TF}} x_{ij}) \text{Ci}(q_{\text{TF}} x_{ij}) + \sin(q_{\text{TF}} x_{ij}) \left(\frac{\pi}{2} - \text{Si}(q_{\text{TF}} x_{ij}) \right) \right], \tag{26}$$

where U_j is a potential height, x_{ij} is a distance between the site i and the j-th line, $q_{\text{TF}} = e^2 k_F/(\pi\varepsilon_0\varepsilon_{\text{r}}\hbar v_F)$ is the Thomas–Fermi wave-vector defined by the electron Fermi velocity $v_F = 3ua/(2\hbar)$ and the Fermi momentum $k_F = \sqrt{\pi|n_e|}$ (related to the electronic carrier density n_e controlled applying the back-gate voltage). Here, $-e < 0$ denotes the electron charge. The Thomas–Fermi wave-vector is also commonly expressed as a function of graphene's structure constant $\alpha_{\text{g}} \equiv e^2 k_F/(4\pi\varepsilon_0\varepsilon_{\text{r}}\hbar v_F)$ according to $q_{\text{TF}} = 4\alpha_{\text{g}} k_F$. We consider two cases: symmetric (attractive and repulsive), $V \gtrless 0$, and asymmetric (repulsive), $V > 0$, potentials, where U_j are chosen randomly in the ranges $[-\triangle, \triangle]$ and $[0, \triangle]$, respectively, with $\triangle$ being the maximal potential height. In order to simplify numerical calculations, we fit the potential (26) by the Lorentzian (Cauchy) function

$$V_i = \sum_{j=1}^{N_{\text{lines}}} U_j \frac{A}{B + Cx_{ij}^2}, \tag{27}$$

where the fitting parameters A, B, C can be calculated from the least-squares method [16]. The typical shapes of the effective potential for both symmetric (attractive–repulsive) and asymmetric (repulsive) cases are illustrated in Fig. 11.

Figure 12 shows the time evolution of the diffusion coefficient within the energy interval $E \in [-0.5u, 0.5u]$ for the symmetric (attractive–repulsive) potential for three different cases of orientation distribution of 50 line defects. (Transport curves for the case of the asymmetric potential exhibit a similar behavior and are not shown here). In the first and the third cases, Figs. 12 (a) and (c), the transport coefficients $D_{\|}$ and $D_{\perp}$ are calculated respectively along and across 50 parallel-oriented lines (distance between them is different and random). In the second case, Fig. 12(b), the lines are randomly distributed, which results in the isotropic transport, i.e. $D_{\text{rnd}} \equiv D_{xx} \equiv D_{yy}$. As expected, the electron transport along the lines are higher than those across the lines, whereas $D_{\perp} < D_{\text{rnd}} < D_{\|}$.

In Figure 13, we show the electron-density dependence of the conductivity of graphene sheets with different (10, 50, 100) number of linear defects for the cases of symmetric and asymmetric potentials. First, for a given defect concentration the conductivity of graphene with the correlated line defects, $\alpha_{\max} = 0°$, increases in comparison to the case of uncorrelated defects, $\alpha_{\max} = 90°$ (see Fig. 13). This can be contrasted with the case of point defects,when the correlation in the defect position practically does not affect the conductivity (Fig. 9). Second, for a given electron density, the relative increase of the conductivity for the case of fully correlated line defects in comparison to the case of uncorrelated ones is higher for a larger defect density. This is an expected result since correlation effect manifests itself stronger for a larger number of objects-to-be-correlated—line defects at hand.

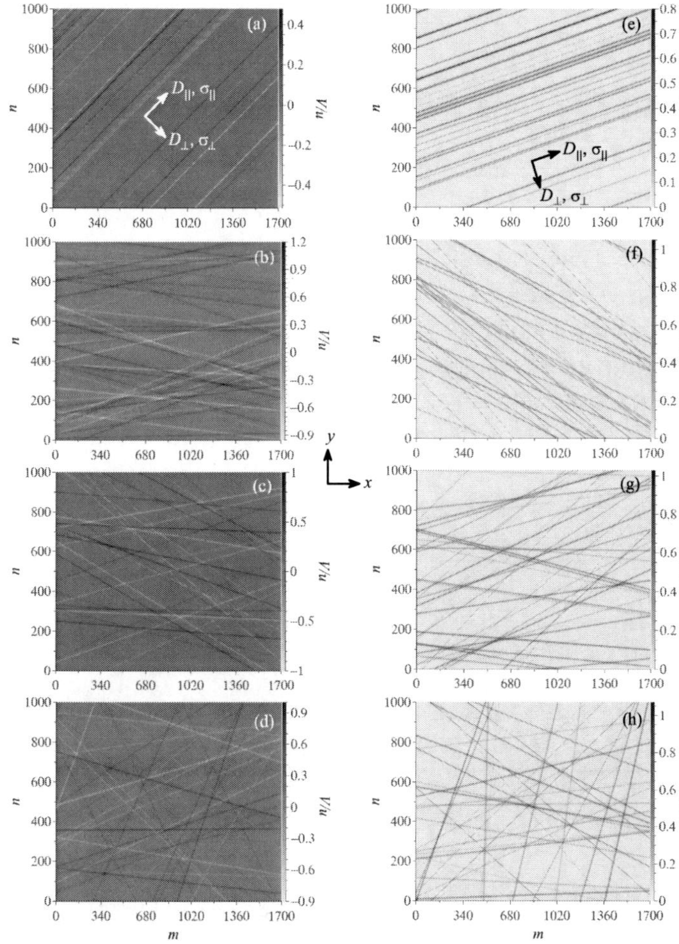

Figure 11. (Color online) Effective symmetric (attractive–repulsive), $V \gtrless 0$, (a)–(d) and asymmetric (repulsive), $V > 0$, (e)–(h) scattering potentials for a representative configuration of 50 orientationally-correlated line defects with different correlation angles $\alpha_{\max}$ (the maximal possible angle between any two lines): 0° (a), (e); 30° (b), (f); 60° (c), (g); 90° (d), (h). Note, that $\alpha_{\max} = 0°$ and $\alpha_{\max} = 90°$ correspond to the cases of parallel and random (totally uncorrelated) lines, respectively. Maximum potential height $\triangle = 0.25u$.

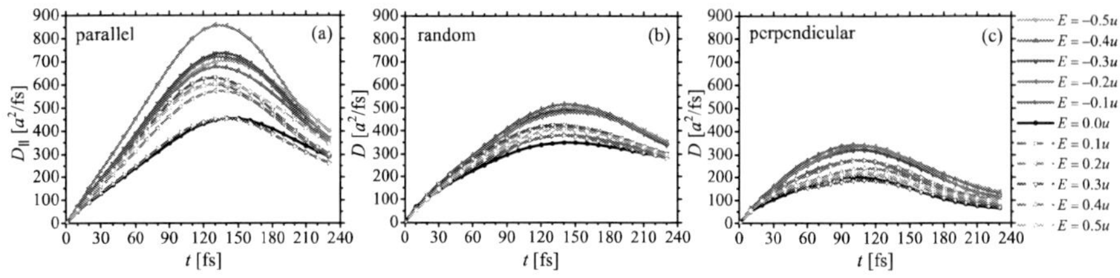

Figure 12. (Color online) Time-dependent electron transport coefficients within the energy interval $E \in [-0.5u, 0.5u]$ for 50 parallel (a), randomly distributed (b), and perpendicular (c) line defects. $D_{\parallel}$ and $D_{\perp}$ denote the transport coefficients in parallel and perpendicular directions to the lines, $D \equiv D_{xx}$ is the transport coefficient along the *x* direction (see Fig. 11). The scattering potential is symmetric ($V \gtrless 0$), the maximal potential height $\triangle = 0.25u$.

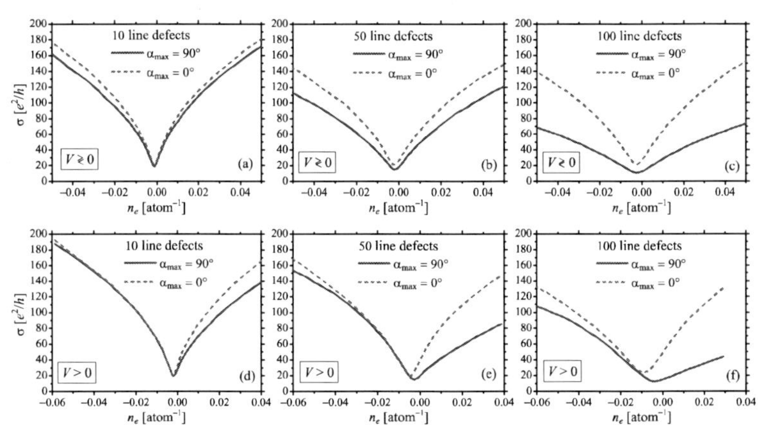

Figure 13. (Color online) Conductivities $\sigma_{xx}^{\alpha_{\max}}$ vs. the relative charge carrier (electron) density for different number (10, 50, 100) of random ($\alpha_{\max} = 90°$) and parallel ($\alpha_{\max} = 0°$) in each realization lines for (a)–(c) symmetric ($V \gtrless 0$) and (d)–(f) asymmetric ($V > 0$) scattering potentials (with $\triangle = 0.25u$). Each curve is averaged over 20 different configurations of lines (including their orientations and distances between them).

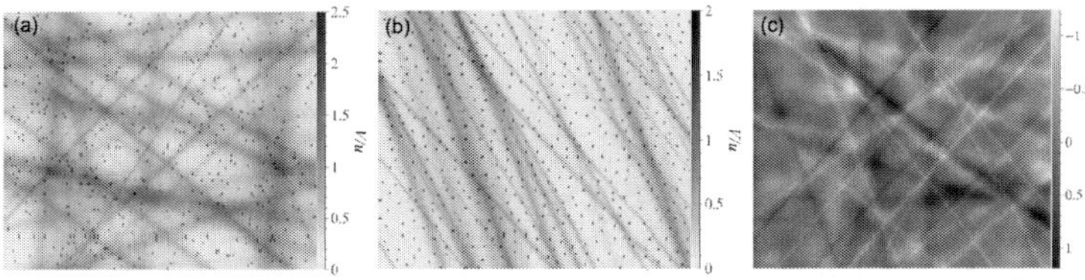

Figure 14. (Color online) Distribution of scattering potential in graphene with a simultaneous presence of random (a), (c) and correlated (b) point and line defects, where the point ones (impurities) are short-range weak (a), (b) and long-range Gaussian (c).

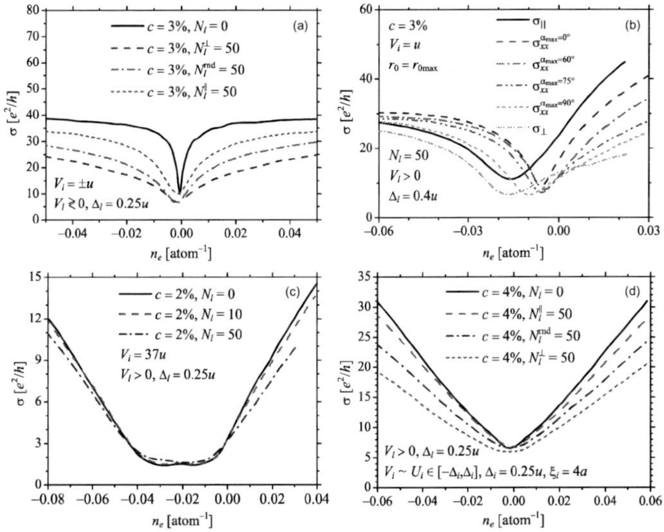

Figure 15. (Color online) Conductivity vs. the electron concentration in graphene with both point and line defects. Subscripts i and l denote impurities and lines, respectively; N is a number of random, parallel ($\|$) or perpendicular ($\perp$) lines. Impurities are short-range weak symmetric (a) or asymmetric (b), short-range strong (c), and long-range Gaussian (d).

Finally, note some features that show the obtained dependencies $\sigma = \sigma(n_e)$ in (CVD and epitaxial) graphene, where the charged line-acting defects are believed to represent the limiting scattering mechanism.

First, the conductivity exhibits a pronounced sublinear electron-density dependence and depends weakly on the Thomas–Fermi screening wavelength [16]. Our numerical calculations are consistent with the recent experimental results [39, 53, 54] that also exhibit sublinear density dependence of σ. This provides an evidence in support that the line defects represent the dominant scattering mechanism in both CVD and epitaxial graphene [39, 44, 53].

Second, the conductivities of samples with different line configurations exhibit significant variations between each other [16]. This is in strong contrast to the case of short- and long-range point scatterers where corresponding conductivities of samples of the same size and impurity concentrations practically did not show any noticeable differences for different impurity configurations [15]. We attribute this to the fact that in contrast to point defects, the line defects are characterized not only by their positions, but also by directions (orientations) and their intersections as well. Such additional characteristics result in much more possible distributions of the potential which, in turn, leads to the differences in the conductivity curves.

Third, for the symmetric potential the conductivity curves are symmetric with respect to the neutrality (Dirac) point, while the asymmetric one leads to the asymmetry in the conductivity, cf. Figs.13(a)–(c) and (d)–(f). Such the asymmetry between the holes and electrons, being also seen in transport calculations in graphene with point [15, 31, 55, 56] and line [16, 44] defects, causes a quantitatively different conductivity enhancements for the orientationally-correlated line defects as Fig. 13 demonstrates.

In conclusion of the section note that the presence of both defect types, point and line ones (Fig. 14), which seems even more realistic than all cases considered above, can significantly affects the behavior of conductivity in comparison with the case when only one type of them is considered (Fig. 15). Conductivity for short-range weak impurities, being electron-density independent, becomes sublinear at the addition of charged line defects to them [Fig. 15(a)]. An interplay between the point and line defects, being modeled by the potential of the same [e.g., positive as in Fig. 15(b)] sign, can suppress the electron–hole asymmetry revealed if they are taken into account separately (see Figs. 9, 10, and

13). However, an addition of the line defects to the short-range strong impurities weakly change the $\sigma = \sigma(n_e)$ dependence [Fig. 15(c)] due to essentially-different scattering forces of the potentials ($V_i \gg V_l$). At last, an addition of the charged line-acting defects to the long-range Gaussian ones, remains the density dependence to be linear [Fig. 15(d)] in spite of its robust sublinear dependence for line defects without point ones.

Effect of Nitrogen or Boron Doping Configurations in Graphene: DFT vs. Kubo–Greenwood Formalism

The present section deals with a comparative implementation of the above-described Kubo–Greenwood formalism and the density-functional-theory-based (DFT) method to calculate the charge transport in a B- or N-doped graphene samples. Boron and nitrogen are the most suitable and therefore commonly used substitutional dopants for incorporation into the graphene lattice. Recent experimental measurements of x-ray photoelectron spectra of N-doped nanotubes and graphene have revealed the presence of several N-doped configurations [57, 58, 59, 60]. Neglecting here the cases of topology changes in the relative positions of the honeycomb-lattice sites (i.e. line defects), there are five configurations for N- or B-doped graphene computational domain. Their geometries are illustrated in Fig. 16. There are one graphite-type defect, N or B substitution in Fig. 16(b), and four pyridine types of defects: dimerized in Fig. 16(c), trimerized in Fig. 16(d), tetramerized in Fig. 16(e), and monomeric in Fig. 16(f). We classify these configurations into two groups: point defects [single dopant atoms (or vacancies), as shown in Fig. 16(b)], and complex ones containing both substitutional impurity atoms and vacancies arranged in a fixed clusters distributed over all structure [Figs. 16(c)–(f)].

Being traditionally a powerful tool for the study of electronic and transport properties of materials, in contrast to the Kubo–Greenwood approach, the DFT method, however, does not allow to treat a large graphene systems. Here, we chose the origin graphene structure (cluster) consisting of 32 sites [Fig. 16(a)] composing a rectangular supercell of the 8.5×9.8 Å size. To exclude the interaction with other graphene sheets, the given supercell is wrapped in a vacuum of 17 Å of thickness along the y-axe. The QUANTUM ESPRESSO computational packet [61] was used for calculations within the electron density func-

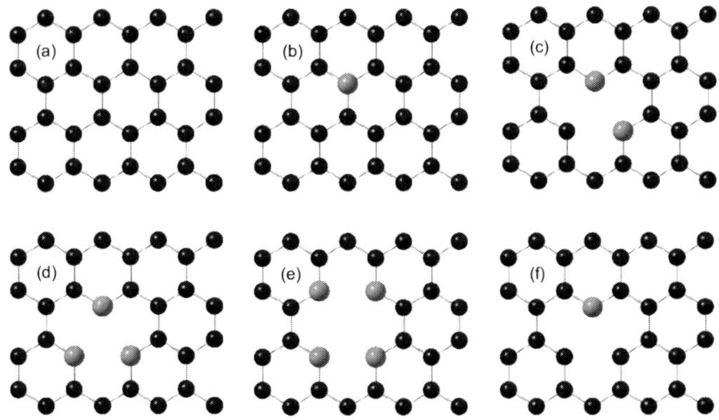

Figure 16. (Color online) Atomic configurations of the pure (a) and N- or B-doped graphene samples with substitutional defect (b), dimerized (c), trimerized (d), tetramerized (e), and monomeric (f) pyridine-type defects.

tional method. To describe the exchange-correlation energy, we used the LDA approach in the Perdew–Zunger parametrization (BLYP for density of states calculation). C, N, and B atoms are described by the corresponding pseudopotentials US-PP [62]. Separation kinetic energy for the wave functions and charge densities are 30 and 300 Ry, respectively. Transport calculations are carried out using the PWCOND codes [63].

As evidenced by the DFT-based calculations of the DOS (Fig. 17), the effect of boron and nitrogen impurities is symmetrical with respect to the Dirac point. Incorporation of boron or nitrogen atoms in substitution within the carbon matrix gives rise to the efficient p- or n-type doping of graphene. Oscillations in DOS for pyridine-type defects [Figs. 17(c) and (d)] become stronger in comparison with DOS for the graphite-type defects [Figs. 17(a) and (b)], that may be associated with a vacancy effect. Note that the oscillations, present in the DOS computed from DFT (Fig. 17), are smoothed in the DOS from the Kubo–Greenwood approach (see DOS for random impurities in Fig. 10) due to the significantly larger computational domain treated within the framework of the Kubo method.

Conductivity in Fig. 18 is calculated within the both DFT and Kubo–Greenwood approach, where in the latter, impurities are modeled by the short-

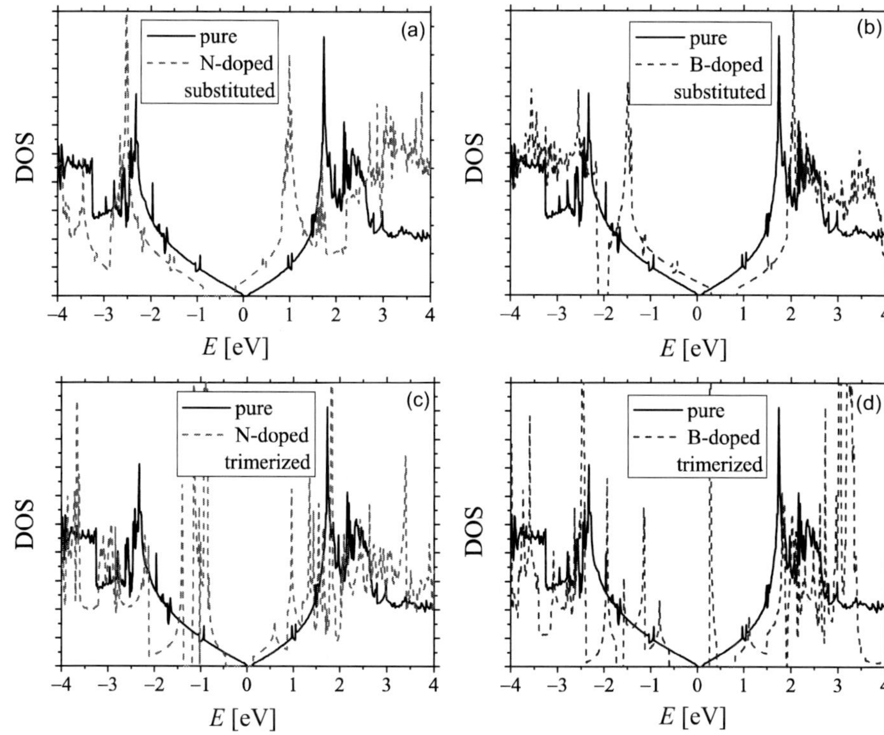

Figure 17. (Color online) Density of states for graphite- (a), (b) and pyridine-like (b), (d) substitutions in N- (a), (c) and B-doped (b), (d) graphene.

range strong scattering potential. The DFT calculations result mainly to the linear energy dependence of the conductivity [Figs. 18(a) and (b)], it means that electron-density dependence $\sigma = \sigma(n_e)$ should be sublinear since $E \propto \sqrt{n_e}$ [15, 16]. However, from the Kubo calculations, $\sigma = \sigma(n_e)$ is obtained to be close to linear [see Figs. 9(c) and 18(c)]. Another difference is that $\sigma_{\min}$ obtained by DFT method is much more smaller than that obtained from the Kubo method. Also, results in Fig. 18(c) are evidence of the fact that the short-range strong impurities manifest themselves as the stronger scatterers as compared with vacancy ones, but this is not the case obtained from DFT [Figs. 18(a), (b)].

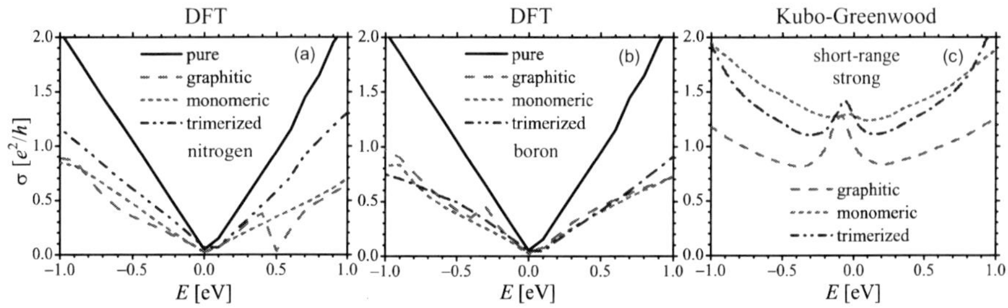

Figure 18. (Color online) Conductivity vs. the Fermi energy calculated within the DFT (a), (b) and Kubo–Greenwood (c) methods for graphene with 12.5% of nitrogen (a), boron (b), and short-range strong impurities (c).

Conclusion

The statistical-thermodynamics and kinetics models of both substitutional and interstitial atomic order in the two-dimensional graphene-based crystal lattices for a wide interval of stoichiometries are constructed. Ordered distributions of substitutional and interstitial atoms over the sites and interstices of the honeycomb lattice at the different compositions and temperatures are predicted and described theoretically. The ranges of values of interatomic-interaction parameters providing the low-temperature superstructural stability are determined within the framework of both the third-nearest-neighbor Ising model and, more realistic, model taking into account interactions of all atoms present in the system at hand. The first model results in the instability of some predicted superstructures, while the second one shows that all predicted superstructures are stable at the certain values of interatomic-interaction energies. Even short-range interatomic interactions provide a stability of some graphene-based superstructures, while only long-range interactions stabilize others. Inasmuch as the intrasublattice and intersublattice interchange (mixing) energies are competitively different with the dominance of the latter, the long-range atomic order parameter(s) may relax to the equilibrium value(s) nonmonotonically.

A numerical study of electronic transport in single-layer graphene is performed by means of an efficient time-dependent real-space Kubo–Greenwood approach, which is especially suited to treat large graphene systems containing millions of atoms. The presence of neutral and/or charged point and/or line defects in graphene is modeled by various short- and long-range scattering potentials. The strong short-range scattering potential describes neutral adatoms covalently bond to graphene. The long-range Gaussian-shaped potential is appropriated for screened charged impurities on graphene and/or dielectric substrate surface. The self-consistent Thomas–Fermi approximation-based effective potential is used for charged line-acting defects (grain boundaries in CVD-grown polycrystalline graphene, atomic substrates in epitaxial graphene, etc).

Correlation in the distribution of impurity atoms gives a slight rise (up to 30%) in the conductivity only for the case of weak short-range potential and only if it is asymmetric (repulsive). In other the most experimentally relevant cases, namely, the short-range strong and long-range Gaussian scatterers, correlation does not affect the conductivity.

Ordering of impurities can give rise to conductivity up to tens times for weak and strong short-range scatterers as compared with the case when dopants are distributed randomly. However, as for the correlation, ordering does not affect the conductivity for the long-range-acting Gaussian potential.

Studying numerically the charge carrier transport in graphene with one-dimensional charged defects, we got electron-density dependencies of the conductivity, which showed some new features as compared with those obtained in case of point defects. First, the conductivity is found to be a robust sublinear function of electronic density and weakly dependent on the Thomas–Fermi screening wavelength. The calculated sublinear density dependence for the case of linear defects is quite different from the case of short- and long-range point scatterers, where the numerical calculations show a density dependence close to linear. We attribute the atypical, but consistent with recent experimental reports, behavior of conductivity to the extended nature of one-dimensional charged defects. Second, the conductivities of samples with different impurity geometries exhibit significant variations between each other. This is due to the fact that in contrast to point defects, the line defects are characterized not only by their positions, but also by directions (orientations) and their intersections as well. Such additional characteristics result in much more possible distributions of the potential which, in turn, leads to the differences in the conductivity curves.

The anisotropy of the conductivity along and across the line defects is revealed, which agrees with the experimental measurements for CVD graphene grown on Cu and epitaxial graphene grown on SiC. For a given concentration of the line defects, the conductivity of graphene with orientationally-correlated defects increases in comparison to the case of the uncorrelated line defects. For a given electron density, a relative increase of the conductivity for the case of fully correlated line defects in comparison to the case of uncorrelated defects is higher for a larger defect density.

A simultaneous account of both point and line defects can qualitatively and quantitatively affect the conductivity behavior in comparison with the case when only one type of them is considered. An interplay between the point and line scatterers modeled by the potential of the same sign suppresses the electron–hole asymmetry revealed if they are taken into account separately. If both point and line defects are correlated and/or ordered, it can give rise in the conductivity of graphene up to hundreds times vs. their random distribution, and thereby can serve as an additional tool to control and govern the transport properties in graphene.

Acknowledgments

T.M.R. benefited immensely from collaboration with Igor Zozoulenko, Artsem Shylau, Aires Ferreira, and appreciates discussions with Stephan Roche and Sergei Sharapov.

References

[1] K. S. Novoselov, A. K. Geim, S. V. Morozov, D. Jiang, Y. Zhang, S. V. Dubonos, I. V. Grigorieva, and A. A. Firsov, *Science* 306, 666 (2004).

[2] K. S. Novoselov, Z. Jiang, Y. Zhang, S. V. Morozov, H. L. Stormer, U. Zeitler, J. C. Maan, G. S. Boebinger, P. Kim, and A. K. Geim, *Science* 317, 1379 (2007).

[3] F. Banhart, J. Kotakovski, and A. Krasheninnikov, *ACS Nano* 5, 26 (2011).

[4] A. H. Castro Neto, F. Guinea, N. M. R. Peres, K. S. Novoselov, and A. K. Geim, *Rev. Mod. Phys.* 81, 109 (2009).

[5] N. M. R. Peres, *Rev. Mod. Phys.* 82, 2673 (2010).

[6] S. Das Sarma, S. Adam, E. H. Hwang, and E. Rossi, *Rev. Mod. Phys.* 83, 407 (2011).

[7] Jun Yan and M. S. Fuhrer, *Phys. Rev. Lett.* 107, 206601 (2011).

[8] Qiuzi Li, E. H. Hwang, E. Rossi, and S. Das Sarma, *Phys. Rev. Lett.* 107, 156601 (2011).

[9] J. W. Klos and I.V. Zozoulenko, *Phys. Rev. B* 82, 081414(R) (2010).

[10] H. Kuramochi, S. Odaka, K. Morita, S. Tanaka, H. Miyazaki, M. V. Lee, S.-L. Li, H. Hiura, and K. Tsukagoshi, *AIP Advances* 2, 012115 (2012).

[11] S. Günther, S. Dänhardt, B. Wang, M.-L. Bocquet, S. Schmitt, and J. Wintterlin, *Nano Lett.* 11, 1895 (2011).

[12] Ch. Held, T. Seyller, and R. Bennewitz, *Beilstein J. Nanotechnol.* 3, 179 (2012).

[13] G.-X. Ni, Yi Zheng, S. Bae, H. Ri Kim, A. Pachoud, Y. S. Kim, C.-L. Tan, D. Im, J.-H. Ahn, B. H. Hong, and B. Özyilmaz, *ACS Nano* 6, 1158 (2012).

[14] M. K. Yakes, D. Gunlycke, J. L. Tedesco, P. M. Campbell, R. L. Myers-Ward, Ch. R. Eddy, Jr., D. K. Gaskill, P. E. Sheehan, and A. R. Laracuente, *Nano Lett.* 10, 1559 (2010).

[15] T. M. Radchenko, A. A. Shylau, and I. V. Zozoulenko, *Phys. Rev. B* 86, 035418 (2012).

[16] T. M. Radchenko, A. A. Shylau, I. V. Zozoulenko, and A. Ferreira, *Phys. Rev. B* 87, 195448 (2013).

[17] A. G. Khachaturyan, *Theory of Structural Transformations in Solids*, Dover Publications, Minola, NY, 2008.

[18] T. M. Radchenko and V. A. Tatarenko, *Nanosistemi, Nanomateriali, Nanotehnologii (Nanosystems, Nanomaterials, Nanotechnologies)* 6, 867 (2008) (in Ukrainian).

[19] T. M. Radchenko, *Metallofizika i Noveishie Tekhnologii* 30, 1021 (2008).

[20] T. M. Radchenko and V. A. Tatarenko, *Solid State Phenomena* 150, 43 (2009).

[21] T. M. Radchenko and V. A. Tatarenko, *Physica E* 42, 2047 (2010).

[22] T. M. Radchenko and V. A. Tatarenko, *Int. J. Hydrogen Energy* 36, 1338 (2010).

[23] T. M. Radchenko and V. A. Tatarenko, *Solid State Sciences* 12, 204 (2010).

[24] I. Yu. Sagalianov, T. M. Radchenko, Yu. I. Prylutskyy, and V. A. Tatarenko, *Uspehi Fiziki Metallov (Progress in Physics of Metals)* 11, 95 (2010) (in Ukrainian).

[25] V. N. Bugaev and V. A. Tatarenko, *Interaction and Arrangement of Atoms in Interstitial Solid Solutions Based on Close-Packed Metals*, Naukova Dumka, Kiev, 1989 (in Russian).

[26] T. M. Radchenko and V. A. Tatarenko, *Nanosistemi, Nanomateriali, Nanotehnologii (Nanosystems, Nanomaterials, Nanotechnologies)* 8, 619 (2010) (in Ukrainian).

[27] D. V. Tuan, J. Kotakoski, T. Louvet, F. Ortmann, J. C. Meyer, and S. Roche, *Nano Lett.* 13, 1730 (2013).

[28] S. Roche, N. Leconte, F. Ortmann, A. Lherbier, D. Soriano, and J.-Ch. Charlier, *Solid State Comm.* 153, 1404 (2012).

[29] T. Markussen, R. Rurali, M. Brandbyge, and A.-P. Jauho, *Phys. Rev. B* 74, 245313 (2006); T. Markussen, Master thesis, Technical University of Denmark, 2006.

[30] S. Yuan, H. De Raedt, and M. I. Katsnelson, *Phys. Rev. B* 82, 115448 (2010).

[31] N. Leconte, A. Lherbier, F. Varchon, P. Ordejon, S. Roche, and J.-C. Charlier, *Phys. Rev. B* 84, 235420 (2011).

[32] A. Lherbier, S. M.-M. Dubois, X. Declerck, Y.-M. Niquet, S. Roche, and J.-Ch. Charlier, *Phys. Rev. B* 86, 075402 (2012).

[33] H. Ishii, N. Kobayashi, and K. Hirose, *Phys. Rev. B* 82, 085435 (2010).

[34] J. P. Robinson, H. Schomerus, L. Oroszlany, and V. I. Fal'ko, *Phys. Rev. Lett.* 101, 196803 (2008).

[35] T. O. Wehling, S. Yuan, A. I. Lichtenstein, A. K. Geim, and M. I. Katsnelson, *Phys. Rev. Lett.* 105, 056802 (2010).

[36] S. Ihnatsenka and G. Kirczenow, *Phys. Rev. B* 83, 245442 (2011).

[37] A. Ferreira, J. Viana-Gomes, J. Nilsson, E. R. Mucciolo, N. M. R. Peres, A. H. Castro Neto, *Phys. Rev. B* 83, 165402 (2011).

[38] O. Madelung, *Introduction to Solid-State Theory*, Springer, Berlin, 1996.

[39] K. S. Kim, Y. Zhao, H. Jang, S. Y. Lee, J. M. Kim, K. S. Kim, J. H. Ahn, P. Kim, J. Y. Choi, and B. H. Hong, *Nature* 457, 706 (2009).

[40] W. A. de Heer, C. Berger, X. Wu, P. N. First, E. H. Conrad, X. Li, T. Li, M. Sprinkle, J. Hass, M. L. Sadowski, M. Potemski, and G. Martinez, *Solid State Comm.* 143, 92 (2007).

[41] B. W. Jeong, J. Ihm, and G.-D. Lee, *Phys. Rev. B* 78, 165403 (2008).

[42] O. V. Yazyev and S. G. Louie, *Phys. Rev. B* 81, 195420 (2010).

[43] S. Malola, H. Hakkinen, and P. Koskinen, *Phys. Rev. B* 81, 165447 (2010).

[44] A. Ferreira, X. Xu, C.-L. Tan, S.-K. Bae, N. M. R. Peres, B.-H. Hong, B. Ozyilmaz, and A. H. Castro Neto, *Eur. Phys. Lett.*, 94, 28003 (2011).

[45] C. Dimitrakopoulos, A. Grill, T. McArdle, Z. Liu, R. Wisnieff, and D. A. Antoniadis, *Appl. Phys. Lett.* 98, 222105 (2011).

[46] S.-H. Ji, J. B. Hannon, R. M. Tromp, V. Perebeinos, J. Tersoff, and F. M. Ross, *Nature Mater.* 11, 114 (2012).

[47] Y. M. Lin, D. B. Farmer, K. A. Jenkins, Y. Wu, J. L. Tedesco, R. L. Myers-Ward, C. R. Eddy, D. K. Gaskill, C. Dimitrakopoulos, and P. Avouris, *IEEE Electron Device Lett.*, 32, 1343 (2011).

[48] T. Low, V. Perebeinos, J. Rersoff, and Ph. Avouris, *Phys. Rev. Lett.* 108, 096601 (2012).

[49] O. V. Yazyev and S. G. Louie, *Nat. Mater.* 9, 806 (2010).

[50] L. Jiang, G. Yu, W. Gao, Z. Liu, and Y. Zheng, *Phys. Rev. B* 86, 165433 (2012)

[51] J. N. B. Rodrigues, N. M. R. Peres, and J. M. B. Lopes dos Santos, *Phys. Rev. B* 86, 214206 (2012).

[52] H. Xu, T. Heinzel, and I. V. Zozoulenko, *Phys. Rev. B* 84, 115409 (2011).

[53] H. S. Song, S. L. Li, H. Miyazaki, S. Sato, K. Hayashi, A. Yamada, N. Yokoyama, and K. Tsukagoshi, *Sci. Rep.* 2, 337 (2012).

[54] A. W. Tsen, L. Brown, M. P. Levendorf, F. Ghahari, P. Y. Huang, R. W. Havener, C. S. Ruiz-Vargas, D. A. Muller, Ph. Kim, and J. Park, *Nature* 467, 305 (2010).

[55] J. P. Robinson, H. Schomerus, L. Oroszlany, and V. I. Fal'ko, *Phys. Rev. Lett.* 101, 196803 (2008).

[56] T. O. Wehling, S. Yuan, A. I. Lichtenstein, A. K. Geim, and M. I. Katsnelson, *Phys. Rev. Lett.* 105, 056802 (2010).

[57] D. Wei, Y. Liu, Y. Wang, H. Zhang, L. Huang, and G. Yu, *Nano Lett.* 9, 1752 (2009).

[58] Y.-C. Lin, C.-Y. Lin, and P.-W. Chiu, *Appl. Phys. Lett.* 96, 133110 (2010).

[59] P. Ayala, A. Grüneis, T. Gemming, D. Grimm, C. Kramberger, M. H. Rümmeli, L. Freire Jr., H. Kuzmany, R. Pfeiffer, A. Barreiro, B. Büchner, and T. Pichler, *J. Phys. Chem. C* 111, 2879 (2007).

[60] B. Guo, Q. Liu, E. Chen, H. Zhu, L. Fang, and J. R. Gong, *Nano Lett.* 10, 4975 (2010).

[61] P. Giannozzi, S. Baroni, N. Bonini, M. Calandra, R. Car, C. Cavazzoni, D. Ceresoli, G.L. Chiarotti, M. Cococcioni, I. Dabo, A. Dal Corso, S. de Gironcoli, S. Fabris, G. Fratesi, R. Gebauer, U. Gerstmann, C. Gougoussis, A. Kokalj, M. Lazzeri, L. Martin-Samos, N. Marzari, F. Mauri, R. Mazzarello, S. Paolini, A. Pasquarello, L. Paulatto, C. Sbraccia, S. Scandolo, G. Sclauzero, A.P. Seitsonen, A. Smogunov, P. Umari, and R. M. Wentzcovitch, *J. Phys.: Condens. Matter.* 21, 395502 (2009).

[62] We used pseudopotentials C.pz-vbc.UPF, B.pz-vbc.UPF, N.pz-vbc.UPF, C.blyp-van_ak.UPF, N.blyp-van_ak.UPF, B.blyp-n-van_ak.UPF at http://www.quantum-espresso.org.

[63] A. Smogunov, A. Dal Corso, and E. Tosatti, *Phys. Rev. B* 70, 045417 (2004).

INDEX

A

B

C

D

E

F

G

H

M

Q

R

S

T

U

V

W

X

Y

Z